IRON BOY

SURVIVING BETA THALASSAEMIA MAJOR

ARTHUR BOZIKAS

Since birth over 8,600 needle sticks,
700 blood transfusions and 2,200 blood packs to stay
alive,
and still counting!

ACKNOWLEDGMENTS

I would like to thank those who helped to make this book possible. Special thanks to my wife, Helen, and our children, Jimmy and Pamela, for their dedicated love and support, and to both our parents and all our immediate family for their tireless affections.

I'd like to dedicate this book to all past, present and future blood donors for their precious generosity. I have a passion for action writing, but their actions have and will always continue to save lives all around the world. I consider them all to be the true action heroes!

Thank you for continuing to DONATE BLOOD!

Arthur

CONTENTS

BOOKS ALSO BY ARTHUR BOZIKAS

THE BOOK GLASSES

BLACK OPS: ZULU. Tom Stiles Thrillers, Book 1

BETA THALASSAEMIA MAJOR

Beta Thalassaemia Major is passed from parent to child in genes. Genes carry information about human characteristics such as eye colour, hair colour and haemoglobin. Haemoglobin disorders vary in their symptoms, ranging from mild to life-threatening. Even within each condition, different people are affected to different degrees.

Haemoglobin disorders include beta thalassaemia major, sickle cell disease, alpha thalassaemia minor and haemoglobin E.

All haemoglobin disorders cause some level of anaemia, which can make you feel tired and drained, and puts extra strain on your heart as it tries to pump oxygen around your body. Thalassaemia is inherited. It is not contagious or transmitted by germs.

People born with beta thalassaemia major are unable to make healthy red blood cells like everyone else and need frequent blood transfusions to survive. In the past, people with beta thalassaemia major did not live past early adulthood. The treatment caused iron to build up in the heart and other organs from all the blood transfusions and that eventually results in heart or organ failure.

This and other related information can be found at the following thalassaemia society links:

- Thalassaemia International Federation https://thalassaemia.org.cy/
- Thalassaemia and Sickle Cell Australia https://www.tasca.org.au/
- Thalassaemia & Sickle Cell Society of NSW https://thalnsw.org.au/

Or contact your closest thalassaemia society near you.

PROLOGUE
OCTOBER 1986

Where was my wife? I found myself sitting alone in front of the obstetrician, who my wife, Helen, and I had met for the first time only minutes earlier.

But now Helen was gone. After another quick look around the doctor's plush office, I excused myself and went out to search for her. Surely she wanted to hear what the doctor had to say?

After dismissing her unusual behaviour as a possible toilet dash, I was stunned to find her outside in the carpark next to our car. She was in hysterics, crying and shaking uncontrollably.

'What's up sweetheart, are you alright? Is the baby okay?'

'I'm not going back in there, I'm not!' Helen screamed through her tears.

'No, of course not, why? What happened honey?' I was confused and now very concerned.

'You know why, you heard him! I'm not terminating our baby!' Helen burst into tears again.

'What?' I said in disbelief. This was news to me. Apparently, I had been more distracted than I'd realised as I had missed the doctor's words.

It was October 1986, I was twenty-five years old, she was twenty-three and this was our first pregnancy. The sense of achievement of being a husband was still new, and soon I was to be a dad. The very thought was overwhelming, wonderful, and scary. It sent me flying high with exhilaration, but then I would remember my medical condition and crash back to earth. The ensuing fear was crushing.

What if our child inherited my condition? Would I be responsible for denying both my wife and our child the chance of a normal life? Panic then manifested inside me, sending my thoughts spiralling out of control. I also now envisioned our baby being born with huge deformities that wasn't even related to my condition.

These misgivings had started the day we'd found out Helen was pregnant. After that, on a daily basis, worst-case scenarios constantly bombarded my mind, exhausting me.

As I'd sat next to her in front of the obstetrician, I'd been assailed yet again by those tumultuous thoughts. Perhaps, if I had reined in my troubled introspection, I would have noticed her leave the room. But overwhelmed by my fears, I was oblivious.

'Look, I can't offer you anything else. Hello, your wife needs you!'

'Sorry, did you say something?'

'I SAID, your wife needs you!' The doctor practically shouting while pointing pretentiously over my shoulder toward the door behind me.

It was twelve when I was first told I wouldn't live past my twenty-fifth birthday. Born with a chronic blood disorder called beta thalassaemia major (Thal), the doctors said my body was unable to produce healthy red blood cells and therefore, I required regular blood transfusions to survive. Since then, it had only been the generosity of wonderful blood donors that had kept me, and others like me, alive. We are known as Thals.

Unfortunately, after so many blood transfusions over the years, all Thals suffer from iron overload and finally succumb to heart and other major organ failures. The doctors said this could happen from childhood through to late teens or, for the lucky ones, early twenties.

Back then, comic books were still the rage. Batman and Superman were always my favourites but every once in a while, I came across an Iron Man comic and was captivated by the hero's suit of armour. The stories were okay, but I was fascinated with what Iron Man could do.

So when I was first told I was going to die from iron overload before my mid-twenties, I wasn't concerned because I truly believed I was Iron Boy. At the time, I felt my older Thal friends didn't know what they were talking about when they said I was going to die because I thought iron made me stronger. I was convinced that I was impenetrable, just like Iron Man. Iron Man comics saved my life at a time when reality would have destroyed my spirit at the most delicate age in my life.

Now I was twenty-five and soon to be a father. I couldn't believe that I was there with Helen visiting an obstetrician, knowing I had reached my shelf life. How could I be so irresponsible, so thoughtless, so stupid?

Then I remembered I was in love with the most beautiful woman in the world.

I may not have been normal, but who is? I knew Helen loved me and that's all I needed to be strong for her and the baby. Then and there, in the doctor's office, I decided to take on whatever challenges this doctor's visit brought, good or bad, because I knew I could achieve anything in life with her love. It was this epiphany that snapped me out of my introspection and made me notice Helen was missing.

But my nightmares were coming true. With my heart racing, and feeling light-headed, I helped Helen into our car while trying to calm her down with a gentle voice. At the same time, my heart was breaking.

'What do you mean the doctor had recommended termination of our baby because of my condition?'

'He said that this baby, being our first, was more likely to have inherited your condition and suggested we should terminate and then try again for a better chance at a healthy baby.'

'Hey, sweetheart, stop crying, he's just an old fart who is stuck in the past. Let's get a second opinion, from another obstetrician. Don't worry about what he said.' But nothing I could say could pacify either of us.

He made this vile and monstrous diagnosis at our very first consultation without taking a blood test from either of us, without consulting with any specialists and without even examining Helen. A doctor's profession is called 'practising medicine' but that advice was unprofessional and unfounded. This advice, from a senior doctor with his experience, a specialist in his field, is something no one would have expected to hear.

Already struggling with a heavy heart regarding my mortality, my worst nightmares were coming true. How was I going to remain strong for Helen?

1

NEW AUSTRALIANS, 1956

The Menzies government era was full of hope and prosperity for Australians. Back then, my mum and dad were both referred to as *new Australians*. Dad arrived in Australia after first starting off from Patras, Greece by ship just before his twenty-first birthday in 1956. His ship was bound for Sydney but ran aground in Melbourne, so he completed his journey by train. He was sponsored to immigrate to Australia by his first cousin, Bill Bourakis who was originally from Corinth, Greece.

My Uncle and Aunty Bourakis already had a beautiful home in Summer Hill, an inner suburb of Sydney. At the time, they were expecting their first child (my cousin John), and my Uncle Bill was eager to show Dad the ropes, to help him find a job and get established in Sydney. Uncle Bill barely knew a dozen English words, which was a great deal more than my dad, who was impressed with Bill's language skills.

My dad, in desperate need of a job, found an advertisement in the newspaper for a machine operator position and told Uncle Bill he could operate any machine, given the chance. This put a smile on Uncle Bill's face, and they set off for the address specified in the ad, an office on the sixth floor of a building in George Street, right in the middle of the city.

On arrival, they found they were the only males answering the advertisement once noticing only women waiting. Uncle Bill assured Dad that this was a good sign. Then he waved the newspaper advertisement under the nose of the receptionist to attract her attention.

'Please job... she for job... she job!' Uncle Bill explained confidently while pointing at Dad at the same time.

Impressed with Uncle Bill's multilingual abilities, Dad stood back, quietly hoping.

The receptionist looked at them with disbelief. 'You know this is for a typewriter machine operator? Is your friend a typewriter machine operator?'

'Yes, oh, yes!' Uncle Bill answered, nodding his head.

The receptionist escorted them to a side office where, to their bewilderment, they discovered a state-of-the-art typewriter sitting on a bare desk in the middle of an empty room. 'Here you go. Tell your friend to show me his skills,' shouted the receptionist rudely.

Uncle Bill looked at Dad and encouraged him to have a go. Without hesitation, he planted himself in the chair and proceeded to type his fingers off. The typewriter sounded like a music box with a ring of a bell every time he started a new line. Uncle Bill and the receptionist watched in wonder as Dad typed away. After completing almost half a page of typing, he sat back, looking confident, and rested his hands behind his head.

With a smile from ear to ear, the receptionist reached over and removed the paper from the typewriter. She stared at the page in bewilderment and then turned the sheet of paper upside down. 'What is this?' she yelled, her smile transforming into a scowl as she shoved the paper in front of Uncle Bill's face.

Uncle Bill looked at it and confidently replied, 'This is Greek!'

'Greek? I want someone who can type Australian, not Greek.' She threw them both out of the office, locking the door behind them.

Jobless and bemused, they exited the building.

Once outside, Uncle Bill turned to Dad. 'When did you learn how to type?'

'I didn't. I saw it in the movies, and I always wanted to try it!' Dad replied wickedly.

———

For the next two years, until my mum arrived in 1958, life for my dad revolved around looking for jobs during the week and playing soccer every afternoon and on weekends. At the height of his playing career, Dad played for Sydney Olympic Football Club. He lived and breathed soccer, dedicating every available minute he had to the sport. Soccer kept him fit and sharpened his mind at a time when he was struggling to adjust to a new life, far away from his family, in a country at the other end of the world.

Yes, soccer wasn't just a pastime for Dad, it was a way of life, seven days a week, for him and for his newfound mates, who were mostly also new Australians.

2

FIRST BORN, 1959

When Mum finally arrived in Australia from Lafka, Greece, also by ship, she was only twenty-one and felt vulnerable after a voyage of almost three months. Away from her home and family for the first time, she felt alone in this strange new country. She even lost contact with new friends she had made along the way almost immediately after they got off the ship in Sydney.

Mum's name was Panawoolla, a name that didn't go down too well in Australia in those early days. Co-workers at the biscuit factory in Camperdown where she got her first job on the conveyor belt would refer to her as 'bloody Nora' because they couldn't remember her name. The name 'Nora' stuck.

When Mum and Dad got married, Mum legally changed her name to Nora. Within a year of her arrival in Australia, Mum and Dad met, fell in love, got married and gave birth to their first son. Nicholas was born on Mum's twenty-second birthday, 1959. He was fortunate to miss out on inheriting Thal, but my parents were soon to discover the truth about our family's so-called 'curse'.

When Nick was born, Dad was working as a labourer and Mum as a cleaner anywhere she could find work, places like large government office buildings in the city and later, an assortment of bus and train depots around the suburbs.

To them, Australia was the lucky country, and people overall were kind and supportive. A large majority of immigrants like my parents settled in the larger cities of Australia, particularly Sydney and Melbourne, contributing significantly to economic and numerical growth. The result of this process has been the development of the nation's multicultural society and my parents were very much a part of this.

In early 1960, they moved to Erskineville, which was home to a large number of immigrants from all over the world, especially from Greece. This made Mum and Dad feel almost at home and was a comfort to them in this trying period of their lives. They were able to comfort one another in their homesickness.

Still adjusting to life in a strange land on the other side of the world, these newlyweds with a newborn baby and a second child on the way were soon given some earth-shattering news.

3

FEBRUARY 1961

My parents welcomed their second son in February 1961, but their joy was soon overshadowed by worry and then horror. I was born at Crown Street Women's Hospital Sydney and, about six months later, was diagnosed with the genetic blood disorder beta thalassaemia major (Thal). This chronic blood disorder meant I wasn't able to produce functioning red blood cells on my own, so I would require regular blood transfusions for the rest of my life to survive.

'Don't worry, your son needs blood, that's all.' The kind nurse would repeat over and over again once the doctor came back a few days later with the diagnoses.

'Blood?' said Mum now confused and relieved.

My parents were devastated. The six months leading up to my diagnosis had been exhausting for them as they'd tried to find what was wrong with me after first noticing my pale and lifeless frail little body. They couldn't understand why their prayers weren't being answered after throwing themselves at the mercy of the various key Greek churches across Sydney.

But after many of the priests they'd approached insisted my parents take me to the Children's Hospital for a check-up, they finally did so. Although the diagnosis was difficult to hear, treatment was started immediately and the relief of seeing me come back to life made their hearts fill with joy again. From then on, they focused on listening to the doctors to understand how they could help me live. They soon arranged for me to have blood transfusions every one to three months or as needed at my parents' discretion.

Following my diagnosis, I was regularly taken to hospital for treatment but being brought up by new Australian parents with strong traditional Greek beliefs, it was drummed into me from an early age that God would look after me if I gave my heart and soul to the Greek Church. Furthermore, I was to be well-mannered at all times and try to do well at school. Manners came easy to me, but school was more difficult considering all the time I missed because of my illness. Likewise, regularly attending the Greek church was something I failed at miserably, and I waited for a punishment that didn't come. It wasn't the Greek church or the presiding Father who bothered me, it was the people attending who did my head in from a very young age.

It wasn't unusual for my mother to take me to attend overnight vigils in the church alongside huge crowds of people. It was like camping but indoors in our best Sunday clothes with other kids my age. I thought staying up all night was awesome.

However, my mother always felt it was important to inform anyone and everyone of my Thal. That triggered brutal comments from most of the adults, especially fellow Greek church patrons, even many times in front of my parents. I recall on many occasions adults would tell my mother and me, 'what the hell are you doing here, take your son away from healthy people!'

Despite enduring their nasty comments, I wasn't a bad kid. I might not have been all good, but I certainly wasn't bad. I decided to show everyone that I wasn't different, that I had the same blood as my brothers, that I was just as healthy as they were and that I was not going to hell. I knew this because I did everything they did, and I gave as good as I got, and my nose bleeds was the same colour as theirs.

During those years and upon repeatedly

hearing these ignorant and hurtful comments, Dad was at a loss about where my condition had come from. The doctors had told them the condition was inherited. In an attempt to make sense of it all, he traced his heritage out loud to Mum, while falling on his knees with a look of failure and despair in his eyes.

'We didn't have this in Patras, where I was born and raised,' Dad said, feeling broken-down, disgraced, and disheartened.

'We didn't have this in my little village of Lafka, either. I have never seen this anywhere,' Mum said in bewilderment, tears flowing freely.

Insults and accusations were hurled back and forth between them until they finally recognised that blaming themselves and each other was futile. It only made things worse. Instead, they realised they should be looking for ways to make me healthier. Eventually, out of their despair came a new-found strength which bonded them.

They discovered that regular blood transfusions allowed me to live a healthy life. It wasn't a cure, but a way to sustain my life at least for a short period of time. But looming ahead were the inevitable and irreversible effects of iron overload on my body and organs, and a prognosis which was beyond belief for my young parents.

The hospital visits started with hopes and

dreams of a quick cure. But their hopes and dreams slowly faded just as their newborn son weakened before their eyes in those early months. Symptoms including paleness, poor appetite, and overwhelming lethargy manifested in increasing severity in the first six months of my life.

With typical Greek superstition, my parents silently feared the worst during those early years, yet always hoped for a miracle. Both puzzled and terrified at the apparent evil that had been inflicted on their second son, they blamed a Greek curse because life was difficult enough without laying blame on either of their families.

At the time, they didn't know they were both carriers of a specific mutated gene. Nor did they know that the chance of a child being born with Thal when both parents carry that gene is one in four. The other three out of four children could be born with alpha thalassaemia (minor) and live an untroubled, normal life, or simply be a carrier.

Our family matched those odds to the letter. My parents had four boys and the only one of the four inherited Thal—me. But despite having to deal with this trauma from the early days of their marriage, my parents showed us nothing other than their unconditional love and devotion. They took what life offered without complaint. To this very day, they are still totally immersed in the lives of their children. In the past, many families have tried to hide the presence of Thal in their family, to avoid the stigma of having a 'curse' in the family.

Once my parents convinced themselves that I was normal, they believed it wholeheartedly. Believing I only needed blood every so often because my body needed a little help and that when people looked down and saw me for the first time as a baby, so pale, weak, and listless, the shocked looks on their faces became etched in their minds. Yet they still wanted to show me off as proud parents do. The hurt and shock may have faded, but it has never disappeared entirely. Even now, at times their eyes will reveal those deep-seated wounds and the fear they had for my life and their family's future.

After the first couple of blood transfusions in that first year I received treatment, my parents were overjoyed to see life seeping back into my weak little body. Their tiny sick baby was soon blooming with healthy colour. Mistaking it for a cure, they were quick to show me off again. Unfortunately, it didn't last long and soon I became pale again and all my original symptoms returned. Their dream of a miraculous cure grew dim and vanished.

Treatment back then wasn't as regimented as it is now. These days we have a quick cross match (blood test) the day before each blood transfusion that now takes about two to three hours per unit of blood to download into each recipient. Things have even progressed for me that I now get a cross match a few hours before my blood transfusion on the very same day. Back then, it was all new to the majority of doctors and specialists. That goes for the entire hospital staff; no one had any answers.

The unfortunate ones who knew what to expect had the difficult task of explaining my prognosis to my parents. On one occasion at a specialist visit, a much older thal's mum approached my mum and told her straight out in Greek that I would continue to have blood for life because there's no cure. They knew I would be unlikely to live beyond my teenage years. My parents were told to come back for more blood transfusions after a ridiculous three- to four-month period or when they thought I looked like I needed blood. But my parents rarely waited that long, and we went back well before then.

My blood transfusion always consisted back then of a stopover in hospital. First, I was required to go in a day or two before the scheduled transfusion for a cross match. Usually, this meant half a day waiting in line at the administration office until I was finally called in for an excruciating finger prick to draw my blood. This sample was then cross-matched with the designated donor's blood of the same blood group for verification and a blood count to determine the exact quantity of blood required. It was usually one unit or approximately a 100–200 ml glass bottle of A-negative blood that was administered, followed by a two or three-night hospital stay for observation.

On occasions, my parents held out longer than the one to three-month period, which resulted in me being carried in barely alive for treatment. This frightened the hell out of my poor parents. These treatments were critical for me because my body was starved of vital nutrients for development.

As a baby, I was fatigued and weak, my skin pale and jaundiced, my abdomen protruded due to an enlarged spleen, my urine was dark and I had abnormally poor bone growth throughout my little body, especially around my upper arms.

The rostered doctor of the day had the most difficult task to find a vein on me for my regular blood transfusion. Failure to find a viable vein resulted in a dreaded 'cut down', when a doctor took a scalpel and cut a vein open, inserting a large and non-flexible needle, compared with today's standard flexible syringe. This was brutally done, and, at times, it felt like it was without any local anaesthetic. These needles were like cold hard steel pipes ripping into your veins with unbearable force.

Unfortunately for me, that wasn't the worst of it because the nightmare concluded with stitches the next day and I'm sure they didn't use local anaesthetics there either.

I considered this to be not only a barbaric prehistoric practice, but it didn't make sense to me because it made that section of the vein unusable afterwards. I couldn't afford to have a growing number of unusable veins. Besides, it required a stitch or two, which hurt more than the cut downs to me.

4

MAYHEM, 1962–65

My brother Nick, older by almost two years, was often puzzled by and envious of the attention I received as a baby. Initially, he was the model big brother, but after two years of watching me receive an inordinate amount of attention, his patience ran out. From then on, he did his best to alert everyone that he was also part of the family. He demanded to be noticed.

Before my brother's reign of mayhem began, I caused some havoc of my own when I developed an appetite for nappy pins. This frightened the wits out of my parents, especially Mum. As if I hadn't given them enough to worry about already!

While the rest of the family prepared to go to church one day, my mother was entertaining an unexpected visitor, a friend of the family. From the lounge room, she heard my panicked cries and quickly ran to check on me. Upon entering my room, she swiftly plucked me out of my cot with one hand. She then held me over the cot and balanced me against her body with my head pointing downwards. As I twisted and turned against her, I vomited profusely. She slipped her finger down my throat and felt a sharp pinprick. Then she examined the clothes she had placed on my bed, which she had planned for me to wear to church. The nappy pin was missing!

Mum's panic turned into sheer terror. She yelled out in Greek, 'My babe has swallowed a nappy pin! Help me, help me! My babe has swallowed a nappy pin!'

The friend's car was outside, and Dad quickly loaded all of us into it and made him drive us to the children's hospital. I passed out during the ride and, upon arrival in the emergency section, they rushed me to get an X-ray. The X-ray clearly showed the open nappy pin lodged in my throat, facing upwards.

Horrified at what she saw on the X-ray, Mum almost passed out. But the doctors calmly explained that there didn't appear to be any damage done and it was a simple procedure to remove it. Before she knew it, they had removed the nappy pin and we were all on our way home, relieved and intact but a little worse for wear. Not surprisingly, we didn't see that poor visitor again for quite a while.

It was after that incident that Nick really kicked into action with many horrific antics. A particularly memorable and perilous episode occurred after he discovered the top-secret location of the family's fireworks stash. Back then, money was scarce, but each year, just in time for cracker night, Mum and Dad brought home a small plastic bag with an assortment of fireworks. To Nick and me, the bag looked huge, and it was always hidden for safekeeping.

Because Mum worked nightshifts, cleaning trains, and Dad worked during the day now as a plasterer and bricklayer, it was difficult for our family to attend any of the bigger fireworks displays in the area. Consequently, our parents made an effort to put together a small display for our family.

On the eve of cracker night, Dad arrived home in plenty of time to see Mum off to work at four p.m. He told us Nick was the babysitter and took off with a man we called Uncle to the nearby club like he has done many times before. Now, Nick had a talent for uncovering secret hiding places, so locating the bag full of fireworks was a walk in the park for him. Determined to get a head start on the cracker night activities, he retrieved a single sky-rocket firecracker from the bag.

Still in broad daylight, Nick placed the pilfered skyrocket on top of the bedcovers on our parents' bed. He opened the bedroom window and aimed the rocket outside. With me and the newest addition to the family, one-year-old Con, standing by his side, he ignited the rocket.

But things didn't go quite as planned. The sky-rocket sat there in full thrust, belching flames. It didn't move an inch. Instead, the bedcovers ignited. Then the rocket exploded with a huge blast and an abundance of coloured flares. The curtains went up in flames too.

I was terrified and found myself quickly out of breath due to inhaling smoke from the burning bed-covers and curtains which were now engulfed in flames. Both Con and I collapsed on the floor, strug-gling to breathe.

In a panic, my idiot brother had another bright idea. An almighty tug brought the curtains down and he quickly rolled them up in the bedcovers. He gathered up the bundle, shoved it in the wardrobe and slammed the door shut, somehow managing to avoid getting burnt.

As the smoke started to dissipate, fearing for my life, I rolled under the bed. How was I to know that was the worst possible place to be in this particular situation?

Barely seven years old, Nick stood there proud of himself for not only putting out the fire but also cleverly hiding the evidence of his misdeeds. Then a huge cloud of smoke poured out of the wardrobe. It was thicker and more sinister now and rapidly filled the bedroom. Spellbound by its sheer intensity, Nick collapsed as his panic turned to exhaustion.

Fortunately, a distressed neighbour called the fire brigade. When they dragged us out of the house, barely breathing, we looked like three drowned kittens. But Mum was furious at Dad for weeks after discovering he left us with Nick and there was no babysitter.

Since that day, I can't recall anyone mentioning fireworks again. But perhaps that's where my fascination with explosions came from, a fascination which would result in me getting third-degree burns on over seventy-five percent of my body many years later when I was in my early teens. But I'm getting ahead of myself.

———

I was about four and a half when the doctors informed my parents that my spleen had to be removed. At the time, we were living in Burwood Road, Concord, a suburb on the outskirts of the city, and I was on one of my regular hospital stopovers. My spleen was terribly enlarged due to my condition, resulting in a distended abdomen. My stomach had also blown up like a balloon, which further complicated things, not only for the doctors but for me as well. My doctors decreed that my spleen needed to be removed immediately.

'The spleen is an organ located below the heart. In an adult, it's a little larger than a tennis ball. It is one of the lymphoid organs, that is, organs where the components of the immune system are produced or stored. The spleen serves as a repository for lymphocytes, the white blood cells that produce antibodies to foreign substances ('antigens'), such as bacteria and viruses, to eliminate these threats.' My doctor explained to my parents' deafened ears.

'The area of the spleen which houses the white blood cells is called the 'white pulp'. After circulating through the lymph vessels throughout the body to patrol for invaders, the white blood cells congregate in the spleen.'

My doctors then turned and tried to explain to my Uncle Tony with my parents eagerly looking on, 'I would put it simply to you, many types of lymphoid cells congregate there, including B lymphocytes, T lymphocytes, and the macrophages that present antigens to them. New red blood cells and other types of blood cells are constantly being produced in the bone marrow, and the spleen also contains 'red pulp', where retired blood cells are broken down. Despite not being necessary for survival, the removal of Arthur's spleen could make him prone to infections. All Thals end up getting their spleen removed sooner or later, but for Arthur, it would be much sooner than later.'

My Thal specialist at the time estimated a short stay in hospital after my surgery, and also said to the three of them that I should be well again in four to six weeks. My mother organised as much time off work as she could, and I was immediately admitted into the children's hospital for splenectomy surgery.

Getting admitted for surgery was like a usual hospital stopover but with more toys. Before my parents could grasp the details of a routine splenectomy, which was anything but routine to them, I was in out of surgery and in the recovery room, minus one poor excuse for a spleen.

'Your son's stomach is large because of the spleen, and it needs removing." The youngest doctor of the group hopelessly failed to explain to my distraught mum and agree to my surgery.

At the time my mum misunderstood this young doctor trying hard to explain the importance of not delaying my surgery in anyway but the more he tried, the more mum believed he was saying that I was dying. It wasn't until an interpreter was located who put my mum at rest, and she agreed to go ahead with the surgery without any further delay.

My recovery time in intensive care stretched from days, into weeks and then months due to complications. The drainage tube which extended from the left side of my stomach to allow excess fluid to drain from the operation site kept leaking fluid well past the estimated timeframe. Mum and Dad were devastated, their fear turning to exhaustion as the days and weeks went by. Yet, despite the circumstances, they were determined to fight for me, and a sense of hope was always found in their eyes.

During this trying period, my mum had a desperate need to do something before time run out for me. The doctors were doing all they could, and she couldn't watch me slip away without doing something herself. Her faith needed to be rekindled. The burden she had placed on herself was so overwhelming and she needed forgiveness for the 'curse'.

Seeking divine compassion, she visited the Greek church in Abbacoby Street, Redfern, where she and Dad had got married. On this day, which coincided with my Greek name day, she went to the church to gather her thoughts and confess her imperfections. She believed she was responsible for our family's curse. Upon her arrival that day, her heart full of despair, she collapsed on her knees under the heavy weight of grief she carried for me.

She prayed that I would survive the complications of my operation. But even if I did pull through on this occasion, the burden of her belief that she had given Thal to me, resulting in a devastatingly short lifespan, was too much for her to accept. Aside from the dire prognosis of my life being cut short, the thought of me repeatedly enduring regular blood transfusions and other complications was beyond comprehension.

With tears streaming down her cheeks and her hands clutched together, on her knees in the middle of this spectacular church, she prayed for absolution. Then she vowed to light a candle in that church on the anniversary of that day for the rest of her life if I was spared from being taken from her at that time. She also vowed she would take me to the cathedral of Our Lady of Tinos, that church's patron saint, on the Island of Tinos in Greece, if I was spared.

After spending the entire afternoon rendered powerless in her desolation, she picked herself up and made her way back to the hospital. She was filled with a sense of relief and a strong anticipation that her prayers would be answered.

It was about three months before I was discharged from hospital and this prolonged crisis took a toll on Mum. After the first month of being by my side almost twenty-four hours a day, she had to return home and juggle her job and home responsibilities, all the while in fear of losing me. After working nightshifts and fulfilling her family commitments at home, she rushed back to my side at the hospital as often as possible. This level of commitment couldn't be sustained and almost put her into a state of insanity.

Meanwhile, Nick was in kindergarten with an enormous attitude emerging, and Dad was doing his best at home managing it all.

When I was released from hospital, Mum was stricken with total exhaustion and ordered to have lots of rest at home or she would need to be admitted to hospital herself. Taking this much deserved rest wasn't too difficult after what I had put her through. Being proud parents was, and still is, the driving force behind their determination, but Mum also knew deep in her heart that her prayer had been answered. I had been spared and she was determined to keep her vow.

5

ADVENTUROUS, 1966

Not long after my splenectomy, we moved to Cope Street, Redfern. This move resulted in an exciting and adventurous period for me. Our new street was full of Greeks, Italians, and other non-English speaking families of European descent and no one could understand anyone else. It was nuts!

Unexpectedly, I found myself with a huge number of friends. It was as if I had an extended family, and we were all in everyone's pockets. On the weekends, the robust aromas of backyard barbecues, freshly cooked bread, and homemade pizzas intermingled and wafted in and out of windows. The volume of TV sets was on high from dawn to dusk and the exciting sounds of Australian wrestling spilled out into the street. Echoes of cage matches between Killer Kowalski, Mario Milano, and the Golden Greek Spiros Arion were heard every weekend. This, of course, fed my obsession for Aussie wrestling, along with that of every other kid in the street, and we mimicked these challengers until they were trodden and crushed.

The front doors of these double storey English-style terrace houses were never closed, with even flyscreens wide open almost any time of the day. Each of the houses was fused to its neighbour. From the top of the street to the bottom, they were lined up on either side of the street like houses in a monopoly game. The individual terrace houses were undistinguishable; different coloured facades were the only way we kids could recognise our own homes.

Starting a new school resulted in a bit of a set-back for my health because I refused to eat lunch or anything my parents put in my lunchbox. Teachers and an assortment of other people, from my older brother Nick through to our neighbours' parents, tried to coax me or force me to eat lunch, with no success. I don't know why I didn't eat. Maybe it was the excitement of starting kindergarten again? Or perhaps the half leg of lamb with Greek cheese and a long roll filled with olives that my mum had stuffed into my lunchbox was too daunting to consider eating? Whatever the reason, it was a hiccup in my transition to a new school.

As a last resort, Mum employed a tactic which embarrassed me beyond belief. She delivered my lunch to me at lunchtime, along with multiple hugs and kisses, right in front of everyone. If that wasn't bad enough, she changed my lunch menu to pies, sausage rolls and Chiko Rolls to make me feel more Aussie. This sunk me to a new level of embarrassment. She also threatened to saddle me with more hugs and kisses if I didn't eat the lunch she brought in.

I decided the food was less embarrassing than the hugs and kisses. That's how my mum finally got me to eat all my lunch, which became an assortment of so-called 'Aussie' sandwiches filled with salami, tomatoes, mortadella, ham, and Greek feta cheese. I was no match for her superior intellect and wit.

Our trips to and from school were exciting in amongst the group of kids from my street, especially with the older ones. We flocked together like sheep yet must have sounded more like congregated cockatoos all talking at the same time. It was a real adventure to walk with my older brother and the older boys.

At the bottom of our street was an orange juice factory, which the older boys sneaked us through to cut our walk to school in half. Even though access was prohibited to the public, the shortcut enticed us daily. So did the dare. Any kid who scored an orange juice on the way through was rewarded with a victory shout and could drink the juice they'd pilfered without sharing it with anyone else. We all thought that anyone with the guts to do that deserved the exclusive enjoyment of their reward.

My opportunity came along unexpectedly one afternoon. By that time, everyone in the group except me had already achieved victory, some of the older boys multiple times over. This left me feeling the pressure of keeping up. But on that dark and wet afternoon, nobody was waiting on the other side of the factory to see me emerge victorious. *Just my luck,* I thought. I didn't even like orange juice.

The real action began after school. The afternoons were filled with billy cart racing and outdoor games to our heart's content. Nick and some of the older boys made up twilight tours just to frighten us younger kids. They commenced immediately after the sun set, just before it was totally dark. The fee was whatever sweets you could get from home. Those with the best sweets were right up front.

The half-demolished double story terrace houses next to the orange juice factory made the perfect backdrop for a ghostly pirate tour. Our tour guides described the demise of the fearless pirate

Blackbeard and how he was buried there. Sound effects were supplied free of charge.

On an evening with a full moon, one of the older boys began with a stage whisper, 'You will see Blackbeard's ghost coming up from the rubble if you look hard enough.' Then while we are focusing in one direction someone out of nowhere would come up behind us all and grab us from behind, frightening the living daylights out of us.

Without fail, that horrifying proposition would send us of all scurrying off home, nearly shitting our pants with fear and excitement. It was scary and preposterous, but we all attended the next full moon tour without hesitation.

————

My normal school experience came to an abrupt end. One day, concerned at why I looked so pale and listless, a teacher picked me up from the middle of the playground and carried me to the first aid room. She gently placed me on the small but comfortable bed and sat down next to me.

She looked in my eyes with deep empathy and asked me with a gentle voice, 'How do you feel now?'

Puzzled by the attention, I remained silent and was bemused because I felt fine.

From a young age whenever I got tired, I knew all I needed to do was to have a quick rest and I would be back up again, recharged, and ready to resume whatever I'd been doing. It was that simple and something I thought everyone did. Answering the teacher's barrage of questions was more tiresome than playing with my friends.

This teacher then got everyone involved from the ladies in the administration office, right through to the principal. It was awful and I wondered what I had done wrong. I was just feeling a little bit tired like I always was.

Before I knew it, my mum was called in, and I was promptly sent home with her. When they asked what was wrong with me, Mum summarised it by stating I had a curse. They didn't understand but it appeared to satisfy their curiosity at the time.

Mum made the most of these incidents, seeing them as a good opportunity to spend more time with me. I couldn't help feeling special every time this happened and didn't dispute the attention.

But Nick would arrive home from school fuming and make things difficult for me because he believed I deliberately tricked everyone into sending me home so I could get more attention. I imagine that due to the combined demands of me and my younger brother Con, who was little more than a baby at the time, he felt all he got were the meagre leftovers of our parents' affection. This started him on an unbelievable course of actions, which mainly were directed towards Con and me.

Con was too young to experience Nick's full wrath, but I quickly learned to be careful to stay away from him, or I risked getting one of his best right hooks across the back of my head. There was much more than that of course, but I don't want to go on about that. Yet sometimes, if he was in a good mood or something went well for him, he was surprisingly supportive, caring, and friendly. If he felt especially generous, he let me play with him, which was very special indeed.

———

There was something different about my next hospital stopover, which made it strange and scary. The trip in on the bus with Mum was the same as usual. Then we waited all day in the administration office before I was whisked away for my usual painful blood crossmatch retrieved via a small pinprick in my finger. What was different about this stopover was that I couldn't remember returning home. In fact, I couldn't remember anything until I woke up about two months later in an isolated sterile hospital ward with Mum by my side.

'Thank you, god, for not taking him just yet.' I

heard my mum saying in Greek, over and over, awaking with a massive headache, feeling sick in the belly, and sweaty all over.

I had somehow acquired a bacterial meningitis infection in my brain which put me out of action for two months. What's more, I had no memory of this time whatsoever. Because of the extreme swelling in my brain, the doctors informed my parents that I had a slim chance of survival. Between the violent sweats and coma-like slumber, the ongoing nightmare was more frightening than a horror movie for my parents. Lucky for me this was while I was unconscious and didn't continue once I awoke.

'Meningitis is an infection of the linings of the brain and ventricles. It's divided into three general categories: pyogenic, granulomatous, and lymphocytic meningitis. Pyogenic (bacterial) meningitis, a potentially life-threatening disease, is an inflammation of the meninges and underlying subarachnoid cerebrospinal fluid (CSF). The specific agents involved vary in different patient age groups and the inflammation may evolve into ventriculitis, empyema, cerebritis and abscess formation. If not treated, bacterial meningitis may lead to lifelong debility or death.' The hospital interpreter explained verbatim the doctor's account to my mum.

'The brain is usually protected from the body's immune system by the meninges, a barrier between the bloodstream and the brain. Normally, this is an advantage since the barrier prevents the body from attacking itself. However, in the case of meningitis, the barrier can become a problem. Once bacteria or other organisms have found their way to the brain, they are isolated from the immune system and can spread. When the body eventually begins to fight the infection, the problem can worsen. As it tries to fight, blood vessels become leaky and allow fluid, white blood cells, and other infection-fighting particles to enter the meninges and brain. This causes swelling and can result in decreasing blood flow to parts of the brain, worsening the symptoms of infection.' Whispered the interpreter, keeping up with the doctor effortlessly.

My mum recalls the doctors looking perplexed after racing in and out several times with different doctors once she alerted them all I was finally awake. Overjoyed at seeing my eyes open, the doctors had to remove her to one side as they focussed on some kind of examination to determine how well I responded. Once they all started to smile and chat with each other in a positive way, Mum knew then I was fine.

My awakening came as somewhat of a surprise to my family, and to an assortment of doctors, specialists, and hospital staff. Still groggy from the effects of the infection and the numerous drugs I had been administered, I recall my parents' astonished faces followed by a barrage of hugs and kisses. I was puzzled by all the attention and bewildered by my surroundings.

Before I knew it, I was whisked off home, blissfully ignorant of the trials I'd endured over the previous two months. Once there, it was back to normal, as if I had only been gone for a few days or so. Nick and the rest of boys carried on with their usual antics and continued leading me into terrifying quandaries, just as they had before I'd got terribly sick.

6

MOVED WEST, 1968

Early in 1968, our family moved west to the corner of Burwood Road and Georges River Road in Croydon Park. It was common in the late 1960s, and the trend persists today, for new Australians to bring their loved ones and family members from overseas to live in Australia. Australia was, and still is, the lucky country and many people like my parents recognised this and immediately wanted their loved ones to share this country with them. They wanted a better life for everyone they cared about. So, with help from my parents, my mum's mother and her siblings moved to Australia from Greece and lived in Cobbitty Avenue, a block away from our new home.

Our new house was next door to a Mobil petrol

station, which made it an exciting place for a young boy to live. I was in wonder most of the time, watching all the different cars coming and going in and out of the petrol station. Sometimes, if I was lucky, I would witness a spectacular car crash, which was mesmerising. We walked past the petrol station every time we went to visit my grandmother and my Uncle Jim and his wife, Anna. Being so close gave us all an opportunity to get to know them better, which was my mum's true agenda. Even though I didn't know it at the time, this bond between us would one day be strengthened during an overseas visit to their territory.

In the meantime, they were in my territory, and I did everything possible to make them feel comfortable. Going to the local corner shop for my grandmother, who didn't speak a word of English, meant I was rewarded with lollies. I also carried out numerous odd paraphrasing tasks and message-taking from business phone calls for my Uncle Jim which kept me occupied and in high demand. I had my own motives for helping him. He had just started up a welding business based in his home garage, and it took my breath away, watching him connect and spot weld steel bars together. He was creative, especially in making steel flyscreen doors, handrails, and a variety of steel balustrades.

On rare occasions, Uncle Jim allowed me into

the garage to watch him work. He put me down in a safe corner and placed a welding mask on me, with strict instructions that I must keep it on at all times while in the garage. I hated having the huge heavy mask strapped firmly to my head. When he first put it on me, the little square window on the front was pitch-black, resulting in total darkness, so I couldn't see a thing. But when he started welding, it was like cracker night with a wonderful array of spectacular sparks and flares shooting in every direction. I could see every single spark and flare as clear as daylight. These times with Uncle Jim were special indeed.

With all this excitement, I didn't have time to miss my friends from Redfern and, what's more, I acquired my first bike. It was a beauty. It was brand new with mirrors, streamers flowing from the gooseneck handlebars, front and rear brakes, whitewall tyres and interchanging gears, and it was painted my favourite colour, red. I was in dreamland, but I didn't yet know how to ride. Nick refused to teach me, under protest because, after seeing my bike, he wanted one himself. His old faithful bike was getting on in years and he was campaigning for an upgrade.

I assume the reason I didn't get hand-me-downs was due to my illness. I remember always feeling special on those occasions, and guilt-free. It didn't matter if it was a wedding or a party, I wore brand-new clothes and shoes and poor Nick had trouser legs almost up to his knees and jacket or shirt sleeves halfway up his arms. It was forever notice-able, but no one said a word. Except for Nick, of course, who complained to our parents that it wasn't fair, but his protests fell on deaf ears.

My new red bike was off-limits to Nick, but he was itching to try it out. The training wheels tem-porarily deterred him, so I made sure they stayed on as long as possible. But not too long after I started riding, I took them off and rode in full flight, with Mum screaming in the background. Dad quickly ran to retrieve me before I killed myself while Nick hung around in his bedroom in protest. Con was still too young to know what was going on.

———

My cousin Arthur Theodosiadis, three years my ju-nior, being the only child in his family at that time, often visited us with his parents. They called us the two Arthurs—he was little Arthur, and I was big Arthur.

Once when they were visiting, Nick insisted little Arthur and I attend a so-called secret meeting with him on the other side of Burwood Road. I wasn't allowed to cross any street, let alone this very busy road, so I refused to go. But Nick seized both little Arthur and me and made us cross the road with him. My heart was racing, and little Arthur kept his eyes closed all away across.

Once across, we quickly ran around to the side of the house directly opposite our house and dropped to all fours to get out of sight. We could hear our parents yelling out for us, but they couldn't see us.

'They wouldn't dare cross the road,' they said in Greek after searching the yard for us.

But we *had* crossed the road. And now we didn't know what to do, so we panicked and sunk further to lie on our bellies! It was about five o'clock on a Sunday afternoon on the summer holidays and we could feel our parents' panic growing as we lay there just across the road from them. Little Arthur was frightened but held back his tears; I guess just being with us made him feel a little less scared. He loved visiting us, but I suspected that Nick could have ruined his chances of visiting us again.

By that time the search was full steam ahead, with the whole family searching frantically for the three of us. Nick made us lie there in total silence, waiting for a break so he could get us home without being spotted.

'Look you guys, we don't have school tomorrow so let's get big Arthur's new bike and all go for a ride. Don't worry, okay?' Nick whispered. There was no response from us, we were scared stiff and looking for direction.

The minutes seemed to drag on like hours and, even though I was excited, I felt it was time to go home so everyone could stop worrying about us. We could hear our mothers' screams escalating as they walked away down the road and then we saw our dads drive off in my Uncle Jim's blue 1965 HR Holden station wagon.

This was the opportunity Nick had been waiting for. He swiftly got us onto our feet and dragged us both back across the road, again with little Arthur's eyes tightly closed. Fortunately, we reached the other side safely, but Nick knew if he didn't get us back into our parents' house quick smart he wouldn't be so lucky. He herded us into the house, only to find our grandmother standing in the kitchen.

'Quickly, quickly, quickly you must leave!' she shouted in Greek. 'If they find you all here they will kill you. You must leave!' she repeated, her arms waving about furiously.

In his panic, Nick decided to carry out his original plan and made me go and get my bike and bring it to the front of the house. Terrified, I complied without hesitation. Nick seized the bike from me and lifted little Arthur onto the gooseneck handlebars. He told me to get on the seat with him, to shut up and hold on. Before I knew it, the three of us were on my new red bike travelling south on Burwood Road towards Campsie with no sense of the consequences.

Poor little Arthur silently held on for his life and I held onto Nick, scared out of my mind. With every push of the pedals, Nick picked up speed and, despite having the three of us on board, he powered down the road with ease. We didn't know where we were going and, for a while, we thought we'd got away without being spotted.

Our fear and excitement had grown to a new level, imbuing us with an unfounded sense of confidence. The afternoon sun shone down on us, and a cool breeze brushed by our faces, reviving us. We were fearlessly interweaving between cars and our excitement turned to exhilaration with shouts of joy coming from both little Arthur and me, while Nick steadily pedalled with all his might.

About ten minutes into our misadventure, my Uncle Jim's blue HR station wagon came out of nowhere. I held on for dear life as Nick's frantic pedalling accelerated us to an unbelievable speed to keep ahead of our pursuers. In a last-ditch attempt to lose my Uncle's car, Nick steered us into oncoming traffic on the other side of the road. Near misses with oncoming cars came too close for comfort but this didn't slow Nick down at all. However, even at this breakneck speed, this cat and mouse game was ultimately futile.

A car manoeuvred sideways from the other side of the road across to our side and blocked our path. In shock and disbelief, I discovered the maniac driver was my Uncle Jim. His spectacular feat of driving forced Nick to swerve off the road and toward a six-foot fence. He locked on the brakes as hard as he could, and we held on for our lives. Just as I nearly lost my grip, our bodies slammed solidly into the fence front on, which luckily cushioned our impact. The next thing I knew, we were covered by a cloud of dust and dirt that concealed all my hurt and humiliation. I lay there at the bottom of the six-foot wire fence in a pile on top of little Arthur and Nick with our legs and arms intertwined.

As the dust began to clear, three familiar faces appeared above us, each with contempt in their eyes. My dad, little Arthur's dad, and Uncle Jim reached out to apprehend us. I braced for the worst of their fury. But both little Arthur and I received surprisingly warm and welcoming hugs from all of them as they carefully picked us up and placed us in the car. Nick, on the other hand, received no sympathy whatsoever, taking the full brunt of their scolding.

A lengthy chastisement ensued before they finally placed Nick and the bike in the car. But to me, the drive home was more terrifying than our crash, listening to my Uncle explaining to Nick what my mum would do to him when she got her hands on him. I sat in the back seat next to my dad in a little round ball full of shame, fearing Mum's response and trying to say a good word about Nick. Little Arthur, however, sat there mutely.

We pulled into our driveway and the car had barely stopped before I was taken into the house by Mum.

Little Arthur went home, and I didn't see him again for a long time. As for Nick? Well, after Mum finished with him, I didn't see him for a week or so either because he was confined to his bedroom. When I finally saw him, his ears still glowed red like a stove's hot plate on full, but he didn't seem fazed by the whole ordeal.

After that, things went on much the same when Nick was on the loose. His next misdeed was stealing milk money from our neighbour's doorstep. I don't know how he came up with these ideas, but they were endless and each one more daring than the last. There was never a dull moment with Nick as a brother and, from an early age, I had no time to be self-absorbed about not being normal.

7

ENFIELD PUBLIC SCHOOL, 1969

Nick was just as provoking and confronting at our local school, Enfield Public. I knew by then to keep my distance from him while we were at school, not only for my safety but for my sanity as well. I don't think Nick knew at the time how destructive he really was. We travelled together to and from school, but as soon as I got there, my friends rescued me from his company.

My invariably poor health and hospital visits took their toll on me and hampered my ability to keep up with my classmates. From the beginning, school meant a great deal to me, but I discovered at an early age that learning was an effort due to all the disruptions in my life. Trying to understand the fundamentals of English and mathematics was an uphill battle. I also lacked much-needed support to complete my homework because Mum said I didn't need schooling, and I was too frightened to ask Nick for help.

One thing I did learn was humiliation and embarrassment in front of my classmates. A girl in my class always sat next to me, not only in the classroom but to my surprise, also in the school playground. I didn't have much to do with her even though we did talk often about school-related issues. Amazingly, she always introduced me to everyone as her friend and she didn't know it at the time, but one day would truly test her friendship.

One morning twenty second-graders, all in full school uniform, proudly wearing our school's logo on our tie and shirt, took our seats in the classroom. I was sitting next to this girl and, being a typical second-grader, totally ignored her.

'Attention, all second-graders, good morning and welcome back to school. Please leave all your books and pens in your school bag because this morning you will all be moving up a grade to grade three,' explained our second-grade teacher, full of excitement for us. She was referred to as 'Miss' and no one dared call her anything else. Miss was firm with no sense of humour whatsoever.

There was a little disruption with a show of enthusiasm from the whole class, but everything settled back down before we knew it.

'Quiet down, quiet down, everyone!' said our teacher firmly. 'Right, this is how it will happen. I will call out each of your names and when you hear your name, stand up and move to the side of the classroom and form a straight line, nice and quietly. Those of you sitting on the right side of the room will stand on the right-hand side and those of you sitting on the left-hand side, please stand on the left-hand side of the classroom.'

As the names were called out, I sat eagerly waiting to hear my name. Miss started with the surnames beginning with 'A' and when she got to 'C' without calling my name out, my heart sank. It further sank as the name of the girl who always sat next to me was called out and she moved to the side of the room.

The seats were emptying fast and all too soon Miss got to 'Z', and I found myself sitting alone in total shame in front of all of my classmates. As I bowed my head in humiliation, Miss congratulated the two lines of students on making it to third grade and promptly ordered them to follow her to their new classroom.

Left alone in the room, I considered getting up and going home but before I could make a move, I heard footsteps heading in my direction. It sounded like a troop of soldiers marching and they were getting closer and closer. I was feeling rather unwell that day and, thinking back, I was well past my transfusion date, so I was weak and as pale as a ghost. My humiliation escalated my situation and I passed out just as Miss returned to the classroom with her troop of new second-graders.

I woke up in the children's hospital with my parents by my bedside and no memory of how I got there. Finding myself in familiar surrounds, I immediately went back to sleep. After a week or so, I eventually went back to school, and it wasn't as bad as I had expected. The little girl from my old class still sat next to me in the school playground and my new classmates turned to me for direction after noticing I knew about everyone and everything. I didn't tell them I was repeating grade two, but my secret was soon discovered, and I finally accepted it, which liberated me to the point that I started concentrating on learning instead of floundering in self-pity.

———

The time finally came when my parents' desire to buy a house became a reality. They knew that renting was getting them nowhere and it was time for them to look out for themselves. They needed to stop sending money overseas. Having helped their families as much as possible, both financially and by relocating them, while trying to raise a family and meeting the additional expenses incurred by my condition, they felt they couldn't let this opportunity pass by.

It was now 1969, and the birth of their fourth son, my brother Angelo, was the big decider for them. It meant moving house again. This time moving was going to be more difficult than ever because they'd move not only away from my grandmother's house but, most importantly, further away from the children's hospital.

When they told us we were moving to Mount Druitt, I imagined us living on top of a mountain. This booming new suburb, almost two hours inland by car, not only offered cheap housing with huge front and backyards but all the houses were brand new.

Moving day arrived with mixed emotions. It was a glorious sunny day, without a cloud in the sky or a breath of a breeze. The removal truck and family car were packed full and the four of us kids were squeezed in together like sardines in the back seat of the car. Trying to balance baby Angelo between the three of us older boys was a constant struggle and the trip to our new home was unpleasant, thanks to the constant pushing and shoving.

Mum packed lunch for us because of how long it would take to get there. We tried to occupy ourselves by keeping a watchful eye on the removal truck with our furniture that was desperately trying to keep up but that didn't work for long. So we bashed each other up until Mum almost lost her voice trying to stop us from killing each other. Battered, bruised and sweaty, we enacted a brief cease-fire when Dad yelled, 'We are almost there. Look, a kangaroo!'

To our astonishment, in a paddock on the left-hand side of the Great Western Highway, not far from our new suburb, sat a dozen or more kangaroos. Taken by surprise, as we gazed out of the car window, we were also amazed to notice that we were not moving onto a mountain. Quite the opposite—there were green farmlands as flat as a pancake stretching as far as the eye could see.

The kangaroos were quickly forgotten and the pushing and shoving in the back seat resumed in full force. Then, we took a right-hand turn and before we knew it, we pulled up in front of our new home. This house and the one next door were the only houses in our street at that time and mud was everywhere. Fortunately, the concrete driveway and paths had already been laid, making it a lot easier for us to move in.

8

HEBERSHAM PUBLIC SCHOOL, 1973

I was still sceptical about this whole moving business as I started slowly moving our things into our new home. However, I was in awe of the house. It was brand new and boasted three bedrooms, polished floorboards, a spectacular kitchen, and a bathroom that was in the house. But the best thing about it was seeing the smiles on my parents' faces, smiles which had long been absent.

The exasperating and exhausting unloading phase took the rest of the afternoon and most of the evening. Dad got takeaway for dinner, a special treat indeed. Then we spread out a pile of blankets on the living room floor and settled in for the night in the middle of the lounge room as if we were camping outdoors.

'Mum and Dad, you said we are starting a new life in our new house, do I still need to get blood now?' I whispered with great optimism.

'No darling, you will always need blood.' My mum didn't hesitate to respond as my dad hugged me even tighter. Amazingly, although I hated the answer, I was comforted at the same time. By the time my three brothers came out of the bathroom our conversation was forgotten from all the screaming and yelling that took place of who was sleeping where.

The next day was devoted to setting up the beds and the rest of the furniture, and none of us kids complained. We were all looking forward to sleeping in our own beds because although camping in the living room was a fun idea, the floor was hard and uncomfortable.

The settling-in period was filled with lots of new adventures and exploring different bike tracks, stuff that was exhilarating for a twelve-year-old. In 1973, I just finished fifth grade and got into sixth grade, and what would become our new local school, Hebersham Public School, was only just built at the end of my street. Until it opened, all the children in our area were forced to attend Blackett Public School, in the adjacent suburb.

The school wasn't too bad, it was the walk to and from Blackett that I detested. It took almost an hour each way. What made it worse were the terrifying attacks from hideous magpies.

A medium-sized magpie is about forty centimetres long and generally black and white. Adult males have a pure white nape and rump, while females are grey in these areas and are slightly smaller. In their breeding season, magpies tend to attack people and animals that wander too close to their nest.

I could usually hear them from a distance, screaming and squawking, and, just when I didn't expect it, a gush of wind swept by the back of my neck. What was even more horrible was that by the time I turned my head to see what had brushed by the magpie would often be back. One particular bird kept swooping on me relentlessly and fearlessly. I was petrified. I curled up on the ground in a ball, motionless and mortified, until someone rescued me.

Usually it was another kid from my school that would come along and save me from the scariest moments in my life. I couldn't understand they weren't afraid. It was astonishing to witness how they would be fearless and stand up to these horrible flying monsters by picking up rocks and throwing it at them. If that didn't work, they would finally pick me up and make me run like the wind until we were out of range.

I still don't know how I survived those traumatic events, but I quickly developed a loathing for magpies and took the long way to and from school to avoid them. After settling into school and familiarising myself with our new suburb by riding everywhere on my bike, the time came to head back to the hospital for a cross match in preparation for my next blood transfusion.

The process was familiar to me by then and I knew I was desperately in need of a blood transfusion at that point. I was apprehensive about travelling to the hospital by public transport from Mount Druitt because it was so far away.

'We could make a day of it and do some shopping on the way home.' Mum assured me the night before that it would be fun, I felt a little better. Dad needed the car, and she was good at making me feel happy with our time together.

I woke up early the next morning, eager for a new adventure, but feeling tired and worn-out due to needing blood. Before we departed for the city, Mum got Nick and Con ready and walked them to school and left Angelo in the care of dad babysitting. I wasn't really conscious of the fact that Mum had worked the night shift cleaning trains at a railway station near central station and that she'd had no sleep at all. She exuded strength and energy as if she'd had the most restful and peaceful night ever. Because she suffered in silence, I didn't give a second thought to her situation.

Off we went, hand in hand, first catching the bus to Mount Druitt Station, with Mum keeping a close eye on my strength and my ability to keep up with her. The train trip into the city was awfully long and sightseeing from these red rattlers was uncomfortable, but still exciting for me, especially travelling with Mum.

First stop for us was Newtown Station, and the rush of commuters flooding out of the train was intense.

'Careful darling, don't worry I'm here with you so take a firm grip of my hand and keep up with me until we get out of the crowd!' Mum instructed me before exiting the train.

I knew if I didn't hold on to her tightly, I could be pushed onto the track or crushed by the crowd by the time we got up the stairs to exit the station. So, as we pulled into the station, I waited by Mum, wide-eyed and determined not to let go of her hand.

The train came to an abrupt halt, the doors flew open, and a flurry of people burst out. It was as if there was a fire on board and everyone was running for their lives. Mum and I headed out of the train onto the platform towards a set of stairs that seemed to stretch up forever. Head down and gripping my mum's hand, I pressed forward with all the strength I could muster.

As we climbed the stairs, the forceful crowd pushed and shoved us. About three quarters up and sweating profusely, I started to draw on my reserve strength prematurely to keep up with Mum. Suddenly, my heart felt like it was exploding with every heartbeat, I felt sick in my stomach and my vision blurred.

The next thing I knew, I woke up in my usual hospital wondering what had happened and how on earth I had got here. Mum had managed to bypass the administration section and the cross-matching area. I was on a ward and in a bed, already tagged, bagged, and half a bottle of blood had been transfused into me. I thought it was great she had managed it all on the same day.

'You passed out on the way up the hideous set of stairs at Newton Station.' Mum explained as I was fading in and out of consciousness.

By the time I was fully conscious, it was much later in the afternoon of that same day. After she ensured I had a meal, she kissed me goodbye, and went home to make dinner for the family and get ready for work that night. I kissed her and looked forward to having a good night's sleep and seeing her again the next morning. A couple of the hospital's finest bottles of A negative were waiting to be transfused into me and I knew it would take most of the night to run through.

I was oblivious to the fact and couldn't even try to contemplate how she did it all because I was either too weak or inept to ever consider the terrible ordeal I put her through on every occasion.

In those days, admittance into a ward for transfusions was a must; it was the hospital's strict rule. The overnight stays often seemed like an eternity, but I fondly remember all the times when they removed the needle from my arm after my transfusion was over, which was always around daybreak. On those rare occasions when I fell asleep, the nurses would wake me to remove my drip. Then, right after they applied a dressing and covered me with the crisp white sheets, the dawn announced a new day, a new beginning. That's how I felt after completing each transfusion at that time of the morning —stronger, brighter, and blessed with a new beginning in my life. It was an awesome feeling and, oddly enough, every time I felt special and lucky to be alive.

———

Before I knew it, the new Hebersham Public School was open for students and sixth grade brought new friendships. My closest and dearest friends at the time were Ron Bertram and Frank Blume, who only lived a spit away. No one could stop us! Everything was ours for the taking. We were all in year six in a brand-new school boasting modern facilities and equipment.

Ron was the most mischievous and funniest of the three of us, while Frank was more serious and less of a risk-taker. Frank came from a stricter background than Ron and me and we did our best to knock the seriousness out of him. I had a lot of other friends around our area but spent most of my time with Ron and Frank. They were awesome and we were always getting up to mischief together.

'Why are you under your house?' I asked with much confusion in the middle of one of the hottest summer's day during the school holidays.

'Who told you I was here?' Ron whispered out from almost total darkness.

'Your mum. Frank, are you there too?' I asked now more confused.

'Shut up and get in here before everyone finds out.' Ron whispered a little louder.

Then when my eyes slowly adjusted as I was gently crawling in I suddenly noticed a couple of girls, the same age as us, with these two giggling uncontrollably. By the time I got all the way in another two girls followed me in and we spent the entire afternoon whispering to each other there. We didn't care how hot, dusty, or filthy we all got in the dirt under Ron's house that afternoon, as long as we had those girls there, everything else wasn't a concern.

Even though we didn't see those girls ever again visiting Ron's place with their parents, Ron kept the tradition up with practically any and every girl in the neighbourhood who would dare to follow there. But Frank and I always knew where to go to find him on those many occasions he would go missing.

9

ROYAL ALEXANDRA HOSPITAL FOR CHILDREN (RAHC), 1973

Located in Camperdown, an inner-western suburb of Sydney, Royal Alexandra Hospital for Children (RAHC) was the only hospital I ever visited during my adolescence. The hospital was relocated to Westmead Children's Hospital in 1996, although the building remains, it has been transformed into contemporary housing units.

By the time I reached my teens, RAHC was a familiar place to me. I had been going there for treatment for as long as I could remember. Being a regular visitor, all the hospital staff knew me well and most of them, from the elevator man to the nurses, treated me as if I were special.

The lead-up to each hospital visit, stopover, and blood transfusion was exhausting. My blood transfusions occurred irregularly, anywhere from one to three or sometimes up to six months between them. Sometimes my parents carried me into the hospital, with me flopped over both their arms because I was too weak to walk. But it was all worth it due to the treatment I received. Usually, my mum took me into the hospital and stayed with me the whole day, every blood transfusion day. She was at my beck and call and gave me small gifts even though money was scarce in our family at the time.

I found myself craving this feeling of being special and started looking forward to my next visit. Things had been going well for a while when, from out of the blue, I received the wonderful news that, at my next blood day, I would be meeting new Thal patients around my age. This left me even more eager and excited than ever.

———

I had just arrived at the hospital when I heard Sister Shaw, whose voice I knew as well as my mum's, call out, 'Arthur!'

Before I had a chance to reply, an unfamiliar deep authoritative voice called out my name. Who could that be? Then Sister Shaw called my name again. She was my guardian angel and had looked after me for as long as I could remember. It was always a nice surprise to see her again and, despite looking forward to meeting the other Thals, the thought of sharing her with them upset me.

She was my power source and, on many occasions, saved me from some of the doctors assigned to treat me, who were horrible. A great number of them came out of the same deep barrels. Besides treating me as if I was not made of flesh and bones, some of them were arrogant and self-centred. They disregarded anything I had to say if I dared to voice my opinion. To them, I wasn't a person and if they felt I needed a cut down after torturing me in an unsuccessful attempt to find a vein, they did so without any care and at times with a great deal of malice.

'Please God, don't let me have a cut down again!' I repeated over and over again to myself as I entered the room where all the IV drips were administered, I found a man there with Sister Shaw.

'Hello, my name is Dr David Bau.' The stranger, who wasn't wearing a white doctor's coat and had his sleeves rolled up, introduced himself to me. He also appeared to be working with Sister Shaw, rather than against her.

'I'm in charge of administering all the IV drips from now on and would like to ask your thoughts on that, is that okay with you?' he explained, in a gentle, authoritative voice. He then requested I introduce myself to him and asked for my advice on a suitable location on my arms to insert the IV drip.

I was taken aback for a second or two with his approach, having learnt from extensive and painful experience to be cautious of doctors. I had a system worked out with Sister Shaw whereby she warned me whether a new doctor was a good one or a bad one. While the new doctor was occupied, she would bob her head up and down if she thought the doctor was good. But, if she didn't move her head at all, I would know to be on guard and expect the worst. On this occasion when I first met Dr Bau, to my joy I saw Sister Shaw's head vigorously bobbing up and down.

Before I knew it, I was crosschecked, tagged, and bagged, without any pain or discomfort whatsoever. I was so fascinated by Dr Bau's approach that I could hardly get a word out. Then I was sent on my merry way, wheeling my IV drip towards my new fake leather armchair. Even though I initially wrote this off as a one-off, it was a sample of what became a wonderful and long-lasting relationship with this super doctor by the name of Dr David Bau.

Until he came along, I didn't know how badly I needed someone like him. He was a godsend and now, I had the perfect team. But I would have to share them both with other Thal patients now.

Still starstruck, I found myself curiously mesmerised, listening to everyone but not able to hear a word. As I gazed around the room and noticed that all the new Thals were stabbed with needles only once by this new doctor, I remember being struck with the thought that Dr Bau's treatment of me was not a one-off occasion and I needed to consider that maybe this doctor was genuine.

Mary and George Lampitsi (brother and sister Thal patients) were the eldest of the group, over four years older than me, Severino (Sev) Scarfo was three years my senior and Peter Karamihalis two years my senior. I was too young to join in with most of their conversations and preferred to sit and stealthily observe everyone.

Before each IV drip was inserted, I usually headed up two floors to the activity centre to pick out an activity to keep me occupied during the long day ahead. The variety of activities on offer was huge—from board games to woodworking items and lots of toys, right through to making my very own leather belt.

It was great to be able to make my own belt, attaching studs and assorted metal bits to it. But now my secret was out, I thought in horror, as Sister Shaw invited all the other Thals up to my activity centre so everyone could select an item to bring back down for the day.

Unfortunately, everyone thought it was a great idea. Due to a shortage of space in the activity centre, we were asked to pair up and had ten minutes to pick something and bring it back down. One of the new Thals, Peter, volunteered to go up with me to pick something for himself. I went up first with my IV drip and he soon caught up to me. His mouth was going ninety miles an hour, all the way up in the elevator. I couldn't get a word in the whole way.

On our arrival, Peter spotted a pair of boxing gloves lying among some sporting equipment which was used by patients who needed physical therapy. He slipped one the gloves on his free hand and, with a devilish stare, tossed me the other glove. 'Here, put this on. I'm not going to hit you!' he said and looked away from me.

Before I had a chance to put the glove on, bang! I copped a massive hit to the head with unanticipated strength and force. Reeling from the blow, I could hardly believe I had just received a king-hit punch from this new Thal patient I'd just met who was wearing one boxing glove.

Dazed and stunned, I stood there, one side of my face inflamed, waiting for his next move. He just stood there opposite me, pleased with himself. I couldn't believe it. He was daring me to don the other glove and fight back. I declined. Picking an activity was now furthest from my mind, so I proceeded back down to the day stay section to continue my transfusion, my head throbbing.

Intending to forget all about this new crowd, especially Peter, I decided to immerse myself in my comic books for the rest of the day. I didn't know it at the time, but the hit I had received was nothing compared to the hit I had coming later in the day. It would turn my world upside down and guarantee my life would never be the same again.

10

IRON BOY, 1973

Comic books were an important part of my blood days. It hadn't always been like that because reading did not come easy to me, as I'd missed a lot of school and I was too sick to catch up at home. Keeping up at school became harder the older I got. The lack of any regimented treatment program for my condition resulted in a reduced quality of life; I constantly felt lethargic and weak.

Sister Shaw had given me my first comic book a few years earlier and although I couldn't read it, I loved it. It was a Batman comic and when she followed it up on my next visit with Superman, I was hooked. My reading skills dramatically improved and by the time she gave me Iron Man, I was able to read quite well.

Iron Man ended up being my favourite. It wasn't because of the stories; it was his suit that mesmerised me. My imagination went into over-drive every time I picked up an Iron Man comic book. I found I wasn't frail anymore. It was Iron Boy who kept bringing me back from the depths of despair and held me together in the midst of so much pain.

From then on, I didn't worry about doctors who couldn't find a vein for my transfusion and would pierce my flesh with hypodermic needles for what seemed liked hours on end without a break and without thinking or caring about how I felt. Even cut downs no longer bothered me, nor the assorted reactions and body shakes I would often get from my blood transfusions. Furthermore, I was now able to deal with being so weak I couldn't even get up and walk to the toilet from my bedroom and was forced to wet my bed. My secret weapon, I imag-ined I was Iron Boy after first hearing I had iron overload from all the blood transfusions, helped me deal with my weakness at a time I needed it the most.

Feeling faint after Peter's punch to my head, I struggled back down two floors to my new armchair, dragging my IV drip with me. I decided to rest up a bit before spilling my guts about what Peter had done to me.

But little got past Sister Shaw. 'What happened to you?' she asked in horror.

Just as I started explaining, Sister Shaw raced out and returned with an icepack and gently placed it on the throbbing side of my face. Then Peter walked in, a huge chess set in one hand, dragging his IV drip stand in the other, still with that idiotic grin of his. What's more, his mouth was still jammed in top gear. But I managed to tell everyone what a brainless idiot he was.

'I gave him a chance to hit me back!' Peter yelled out loud and laughed it off, as if it were a big joke.

'Peter, there will be none of that here. You are older and should know better. Don't let me catch you doing that again, are you listening to me?' Sister Shaw went off at him that was a joy to watch but he just ignored her.

To my surprise, George and Mary as well as Sev already held Peter in contempt. They sided with me, which instantly made me feel a lot better. Peter continued to mouth off about nothing and tried to make it up to me by asking me to play a game of chess with him. I declined, choosing to rest quietly and recover before my mum saw me like that.

Lunch was a treat at the children's hospital back then, with generous servings of roast chicken, pork, beef, or lamb with real gravy and steaming hot vegetables and mashed potatoes, finished off with a huge variety of desserts and drinks. Thinking back now, I can't believe that we would all sneer at the sight of the lunch trolley and dole out comments like, 'What shit are we being dished up today? I'm not hungry. I'll have everything but the vegetables!'

We didn't know how well off we were, getting lunch served to us on a hot plate by people who cared about us. On this particular day, we all managed to avoid making much of a fuss and quietly finished off our lunch without a single complaint.

After lunch, we were all on our last bottle of blood; all the blood was stored in glass bottles back then so extra care needed to be taken. It was amazing to see all of our IV drips going simultaneously and Sister Shaw was incredible, looking after all of us. It was exhilarating and scary, but I enjoyed the excitement.

By then Peter had challenged everyone there to a game of chess and mostly lost, so he turned his attention back to me. 'Um... Arthur, tell me, what do you know about thalassemia?' Before I had a chance to respond he continued, 'Did you know you won't live to reach adulthood? You know you're going to be dead before your early twenties?'

'Shut up, Peter! Don't listen to him, Arthur. No one listens to him. Why are you starting on that, you dickhead? Shut up!' said Sev.

'I'm trying to get him used to it. It's not my fault he doesn't know the truth,' Peter said putting his nose in the air.

'Shut up, Peter!' said George.

Peter laughed. 'Well, he'd better get used to it, because he hasn't got that long to go before he kicks it, if you know what I mean?'

'Shut up!' shouted everyone in the room in chorus, which did the trick and temporally silenced the dimwit.

While amused by our group's coordinated efforts, I was shattered by what Peter had said. An overwhelming flood of questions ran through my head. Overcome with mixed emotions, my heart was racing, and I was finding it hard to breathe. Then, slowly, I started to compose myself and gather my scattered thoughts. I reached for my safety blanket, my Iron Man comic books, desperately trying to block out his words, which were still hanging in the silence of the room, words that could not be unsaid. But I couldn't focus on the comic. I needed to know more.

'I haven't got an illness. I just need blood like you all do,' I said.

'Sure, Arthur. Whatever you say. Don't blame me that I didn't warn you,' said Peter with that stupid grin plastered on his face again.

After two or three minutes of silence, everyone

carried on as normal with an assortment of conversations, albeit restricted to safe topics. I went back to my reading.

Soon, my mum returned from the shops, balancing a few bags in each hand, and looking pleased to see me among my new mates.

'Come over here, Mrs Bozikas, and have a cold drink. Arthur hasn't long to go before he is finished, about thirty minutes now,' said Sister Shaw, lending a hand with Mum's shopping.

My frustration grew—I needed my questions answered immediately. I tried to get Mum's attention without all the others noticing. My efforts were in vain, so I opted to relinquish any further attempts until we were alone.

Finally, I found myself alone with my mum in the adjacent room getting ready for my IV drip to be removed from my arm. My bottle of A-negative blood was almost completely empty and as Mum and I watched the last of it leave the bottle and slither down into my IV tube, I quickly asked, 'Mum, what's thalassemia? Am I going to die before I get too old?'

Before she could reply, Doctor Bau and Sister Shaw returned to remove my IV drip and then sent us on our merry way home. Glad this day was over, I couldn't believe I was the last to leave.

As we left the hospital that afternoon, I was determined not to bring up the subject of thalassemia ever again. My face was still throbbing from Peter's punch earlier that morning, but I concealed it from Mum, and she didn't notice. She was happy I'd found some new friends my age who could keep me company while I was getting my transfusions.

On the way home in the car, I filled her in about the other Thal kids and especially about Peter. I told her what I thought of Peter and his stupid grin but kept quiet about his punch. I thought Mary and George were great and that Sev was too much older than me for us to become good friends, but he was okay.

The drive home was over an hour and a half long and, just before we got home, Mum asked a strange question, 'Why have you talked about Peter almost all the way home if you hated him the most?'

Baffled by her observation, I was speechless. 'I don't know!'

'Look, give him another chance next time before you make up your mind about him,' said Mum, half in Greek.

'No way. I'm going to kick his arse next time I see him!' I soothed my sore face with the palm of my hand, being careful to conceal my actions.

11

A HIGH SCHOOL TRADITION, 1975

In 1975, I was thirteen and managed to finally get to high school. My first-day nerves were magnified by the terrible rumours floating around at the time about how badly new students were treated by the older students at Mount Druitt High School.

Before entering the school grounds, I noticed a large number of hippies and sharpies in the school population. The rest of the crowd leant towards a more contemporary modern appearance which I knew after looking more closely at them reflected conservative attitudes and views. I immediately integrated myself into the latter group and never drifted away.

As for the hippies and sharpies, they could keep

all that shit to themselves. I didn't want to go down that track. Even though there was nothing wrong with being a hippie or a sharpie because they were just expressing themselves, I disliked the fashion and the music.

I was apprehensive about anyone finding out about my Thal besides my closest friends, Ron and Frank. In fact, I often rubbed my pale cheeks as hard as I could to get blood circulating through my face, so I didn't look sick. I didn't want anyone to know because I was convinced I wasn't sick! I was normal and attending a normal high school and no one was going to tell me otherwise. I just needed some blood every so often, that's all.

While I was trying to convince myself that was true, the demons in the back of my mind shouted Peter's words that I was going to die before reaching adulthood. This frightened the life out of me, stopped me in my tracks, and sucked the breath from my lungs. On these occasions, I took a long deep breath to calm myself and continued on my way as if nothing had happened, so no one knew I was worried.

Mount Druitt High had a school tradition at the time that new students would get tagged on their first day. Unfortunate individuals who resisted being tagged would receive an atomic wedgie, which was getting your undies pulled up over your head where you were standing. This tradition was not worthy of the school's academic achievements, I'm sure.

Anyway, I figured that having the label on the back of my school shirt viciously ripped off and being tagged in the middle of the school grounds was preferable to getting an atomic wedgie. So the first thing I did was head straight for the middle of the school grounds and wait for the inevitable.

Surrounded by the evidence of prior skirmishes targeting my new classmates with shirt tags and hideous torn scraps of undies spread all around, I waited. It was horrible but, like a soldier fighting for flag and family, I knew what I had to do. Bracing myself for the worst, I stood there with my head held high. Tension built as I watched my fellow students falling one by one before my eyes. The carnage was overwhelming and, just when I thought I couldn't bear it any longer, I was violently grabbed by the sleeve of my school shirt. *This is it*, I thought.

'What the hell are you doing here? Get out of sight, you stupid idiot!' a familiar voice screamed.

Shocked and surprised, I turned to face my older brother, Nick, who was in year eleven and a senior at the school. Trying desperately to control my panic, I stood there speechless. For the first time in my life, I was relieved to see him.

Before I could reply, a group of senior boys

rushed towards us like a pack of hungry wolves. Nick and I found ourselves surrounded by eight to ten giants, each of whom, at almost six foot tall, towered over us. To my disbelief, Nick strictly instructed those huge seniors that I wasn't to be touched. Without any argument whatsoever and with fear in their eyes, they all turned away, leaving Nick and me alone.

In contrast to the departing giants, Nick stood at just slightly shorter than me and yet they listened to him as if he were a teacher. Stunned, I stood there mutely as Nick talked to me about who knows what; I didn't take in a word he said. Then he left me alone, feeling confused about what had just happened.

When the school bell rang, I gathered up my composure and staggered off like a drunk to my first class. *I can't believe that happened,* I thought to myself over and over again.

By the time I got to class, word had got around the school that I wasn't to be touched because I was Nick's little brother. From that day on, I had a newfound respect for Nick. I was astounded at how much power he had at this school but most of all, I was spared total humiliation in front of my peers.

Nick's protection somehow prompted a show of respect for me as well, which propelled me into unfamiliar territory. From my first day of high school, I found myself with heaps of friends and an unaccustomed sense of freedom. I was amazed at my sudden popularity and clout, all because of Nick. It was weird that we got on wonderfully at high school, while at home it was World War III.

12

LONG-DISTANCE RACE, 1975

On rare occasions, if she couldn't find a babysitter, Mum brought all my brothers to the hospital with us, but only on my blood cross-matching days. She wouldn't do this often because they were more than a handful. They tore away from her in all directions and she and I found them in the weirdest places. Nick was always found in the elevator, going up and down, but the other two were often found in the canteen or under a hospital bed in a vacant ward playing camping out.

In my teens, I made it clear to Mum that I didn't want them to come along. They didn't understand what I was going through; not even Nick grasped my situation. Instead, they looked upon it as a day out or an adventure. It took me over six months to convince her to let me go in alone, with Peter motivating me with insults.

Initially, I was allowed to go in alone only on my cross-matching days, but before long, I was going in alone on my blood days too. Dad picked me up at the end of the day after I'd been transfused with a couple of glass bottles of the hospital's finest A-negative blood, alongside my Thal mates.

Going in alone made me feel as if I were normal, like my new-found Thal mates, not yet truly grasping the dreadfulness of what my illness had in store for me.

Around the same time, Ron and Frank thought it was unfair I had an excellent excuse to get out of a lot of school, but they were oddly kind about my condition. Ron's mum and dad, who were of Scottish descent, decided to move to England the following year. The whole family was going—Ron, his two younger brothers and their little sister. Frank and I could hardly imagine not having Ron around. We tried to convince his parents to stay but our efforts were futile. We knew that the energy and laughter Ron brought into our lives would be greatly missed when he left, so we made an effort to make the last year unforgettable.

Ron had the knack of turning a difficult situation into something funny and spectacular. He was brilliant with people, especially with girls. He was the shortest of the three of us when we were growing up but did things that made him look and seem the tallest of all of us.

On one of many similar occasions Ron pulled the pants down of a boy who tried to pick a fight with him during a game of basketball lunch time at school in front of all the girls that were watching from the sideline. The guy was a monster but after doing that to him he didn't go near Ron because the girls would be following Ron to see what mockeries he would do next.

From a young age, he was interested in anything to do with aviation, so he managed to get the two of us into the Australian Air League (AAL), which was awesome. AAL was a youth organisation like boy scouts but with an aviation angle. It was held for a few hours on a Friday night in the high school gym auditorium in a nearby suburb because it was a new youth group to the area.

AAL meetings was mostly learning about all sort of disciplines like marching, respect, teamwork, and skills relating to improving your knowledge with all things related to aviation. Regular parades and other events like camping are often organised throughout the year but workshops on learning about flying and aircrafts consumed most of the weekly meetings.

Meanwhile, Frank was attending a scout group

because that's what his father preferred. Frank had a strict upbringing, with both his parents being of German descent. His father ruled the house with an iron fist.

Frank, Ron, and I had always been in the same class at primary school and when we got to high school, Frank's parents divorced. He stayed in the family home with his mother, while some of his siblings went and lived with their dad and the others were old enough to live elsewhere. The three of us were always treated well when Ron and I visited Frank's place, both before and after his parents' separation.

Ron's serious side came out while at home and when attending the AAL. It was strange to see him behaving so well in front of his parents and at the AAL, whereas with us he was frantic and unpredictable. Ron's parents were also great to me and Frank, but we couldn't understand why Ron was so well behaved at home.

One of the many challenges between Ron and I was to see who would achieve the highest rank at the AAL. Early on, I figured that, with Ron's attitude and behaviour, he had no chance of rising through the ranks. But before long, he went up to new cadets and lectured them about their misbehaviour. The gloves came off then and the challenge became serious, but I knew deep down I didn't stand a chance because my health and stamina were always getting in the way. In fewer than two years, Ron outranked me and won more trophies than me.

The AAL was the best experience I could wish for at the time I needed it the most, which was just before going to high school. It taught me responsibility, respect, honour and to believe in myself at an early age that gave me a fundamental base to build on, all thanks to Ron. But I was still hopeless at sport.

During our first year of high school, every sporting activity Ron tried out for he was accepted into, even basketball, despite his short stature. On the other hand, I failed at everything, but that didn't stop me from trying. From as early as grade five, I knew I was hopeless at sport when I failed to make it into the school baseball team. Ron and Frank got on the team, and I was left behind the baseline watching them. I wasn't even allowed to collect the balls, that's how shithouse I was at sports.

My moment finally came at almost the end of the first year of high school at the annual long-distance running event. The cross-country course was hard and long. The two hours estimated to complete the event was too ambitious for me and I didn't know what I was thinking. It was an event of historical importance for the school, and they took great pride in holding this event.

'Hey where are you going?' A year twelve student said with an official school sports vest on pointing at me.

'I'm in year seven, I don't think I can do this?' I replied not thinking what to say.

'All students from year seven right through to year ten that's everyone must take part no matter what issue or complaint they have, so get back into line now.' It was made loud and clear to me for all to hear.

Upon hearing this, my heart dropped to the floor, and I knew there was no escaping it. The seniors coordinated the event and the teachers supervised it. They had no problems if students wanted to walk the whole way, as long as they participated.

The time finally arrived and Ron, Frank and I headed for the starting area without saying a word while being watched. They both felt my apprehension about the event and didn't know what to say or do about it, trying hard not to look at me because they could see I had a terrified look on my face. After picking up on their discomfort, I told them that I was going to kick their arses and beat them both hands down. But they knew me too well and didn't buy into it for a minute.

The next thing I knew, we were in the midst of

a crowd of students waiting for the siren to go off. There were hundreds of students, both girls and boys, and the excitement and nervous energy were almost tangible. It was an awesome feeling, and I couldn't help getting caught up in the moment. Then a loud belch sounded, which was the siren, and everyone ran for their lives, including me.

Right away, I lost track of both Ron and Frank as I made my way in the middle of the crowd at a slow pace. After a minute or two running at an even slower pace, while frantically looking around for my friends, I came across the first obstacle in the race—an empty creek bed with steep banks on both sides. Without thinking about it, I negotiated my way across effortlessly and I had a revelation—I wasn't too bad at this cross country running.

While cutting across the creek, I passed thirty to forty fellow students with ease, so I pressed on and picked up my pace a little. Senior students were lined up along the side of the path and drink tables were positioned about fifty metres apart. I found myself entering another suburb and, surprisingly, I was still not feeling at all tired. Just ahead of me a senior student extended his hand to both show me the way and offer me a cup of water. I grabbed the cup, quickly drank the water, and tossed the empty cup in a bin few metres on, all without stopping.

As I kept on passing other students along the way, I had an unstoppable desire to go faster. I didn't know my limits because I'd never run in a race, and I didn't know how far to push myself. With still no sign of Ron and Frank, after running steadily for almost an hour, I decided to go for it and break away from my long rhythmic stride into a faster pace to see what would happen.

It was an awesome feeling and a newfound high for me. My heart was beating as fast as it could and a cool breeze swept across my face, instantly cooling me down. My legs started to complain, and my pace began to slow, as my breath became more laboured. I thought about stopping and collapsing on the grass. I couldn't believe that this stupid race was so long, but I was still overtaking other students.

Then, just in the nick of time, I was back on the school grounds, heading towards the finishing line. Ron and Frank were still nowhere in sight. I was almost completely out of breath when I crossed the line, coming in an impressive thirty-seventh place of the over 300 students who competed in the long-distance race.

The great thing about the event was that my brother Nick was one of the officials at the finishing line and saw me cross the line. He stood there with his mouth open, and I could tell he was really shocked. Standing next to Nick was Ron, who'd come in tenth. As for Frank, well, he was too busy picking up girls and managed to walk with two beautiful students all the way around.

It ended up a great day because not only did I finish the race, but I also discovered I wasn't as bad at sports as I'd always thought I was. But more importantly, I discovered a little bit about myself. I don't know why I did so well on the day, but if I hadn't been forced to take part, I would not have realised that it wasn't my illness that disadvantaged me, it was my outlook and attitude.

The penny finally dropped for me on that day, not only about accepting my condition, but I learnt that I needed to concentrate on my abilities. Also, I came to realise that by knowing my weaknesses, I could turn them into strengths, which liberated me enormously. It changed my whole thinking about being an individual and accepting who I was. All I had to do now was to persuade my parents, which was much harder than playing any sports or taking part in any activity at home or school.

My Thal mates at the hospital all knew about this already and had a great outlook on Thal. Their attitude toward each blood day was that it was a day to get together or, to put it simply, a day off school. Spending time with them was liberating for me. I had a huge thirst for knowledge and on every possible occasion, I encouraged them to tell me more about all the different activities they did.

Severino hardly ever responded to my curiosity, but Peter did enough talking for all five of us. Mary and George were supportive and tried to answer most of my questions, while Peter kept reminding me that I was going to die before I was an adult.

These reminders about our mortality always prompted a volley of shouts and shut ups from all of us, including Severino, all aimed at Peter. Unfortunately, that didn't stop him—he kept on repeating it. Mary, my favourite of the group, really gave it to him and was the most protective. We didn't know at the time that we weren't going to have her with us for long, as she would succumb to iron overload only a few years later, but while she was with us, she was protective of all of us, including Peter.

Peter didn't care who was in the room when he opened his big mouth. Sometimes my parents and family members would drop in to see how I was going, and they too were subjected to Peter's uncensored comments. I hated it when my mother heard what Peter said, as afterwards, she would cry, realising that with no treatment available, there was some truth to his words. I would also get upset every time he said it, but I would try hard not to show it in front of Mum.

On those rare occasions when my Uncles and

aunties visited me at the children's hospital while I was in the company of my Thal mates, it wasn't Peter's big mouth that upset me the most. I couldn't help feeling I was the abnormal sick boy in the family they all came in to see. They stopped by at the hospital to visit me, dressed immaculately on the way to a day's outing somewhere. I knew they meant well, and they all loved me very much, but I was with my Thal mates, unaccompanied by my parents. I couldn't understand it at the time, as they could have visited me at home with my parents there, so I felt like I was a freak put on display.

After each short, odd visit, I realised that sometimes people don't think. I also wondered if maybe it was me who had the issues and not my Uncles and Aunties. After all, they took the time to come and see me. It was all too confusing dealing with it all.

I reflected that I needed to focus on the positives, as I was constantly placing my small achievements to one side. Instead of feeding my negative thoughts, I should have been grateful that they took the time out of their day to visit me. Acknowledging this kept me on the right track. The long-distance race of this illness was not over for me, and although I was aware my race might be cut short, I was more prepared for this particular endurance session than ever before.

13

THIRD-DEGREE BURNS, 1976

Towards the end of 1976, Ron and his family left for England, which devastated me. Walking to his house was a daily routine and all of a sudden he was gone. In the beginning, I continued to walk past his place but that didn't help fill the void.

Even though we all made it through the first few years at high school unscarred, Ron had been the most popular one by far. No matter what the challenge, the sport or classroom activity, Ron was somehow brilliant at it. At less than five feet tall and as skinny as a broomstick, it was hard to believe he could be that good. When he got into a fistfight, to get a good punch in, he first had to jump up and lean on his opponent's shoulder. Nothing stopped him! He was also the guy slamming the points at

the school basketball match or answering that difficult question in class, no matter whether he got it
right or wrong. Frank called it pure luck, but that's
how Ron was.

Ron and Frank knew all about my Thal from
primary school days and they were fine with it.
After I'd first told them, Ron kept telling me that a
vampire was sucking out all my blood while I was
asleep.

'Why do you get blood again?' Ron would often
scream it out for all to hear spontaneously and always at an inappropriate time.

'He's only joking!' I would always reply back
with a half-smile and a death stared directed at him.

I obviously didn't believe him about the vampire comment... after a few sleepless nights.

Frank was quiet about my Thal at the start and
hardly mentioned it again. But after Ron left, he
surprised me by asking me if he could go along with
me on my next blood day. He referred to it as
having a day off school too, but I think he was
telling me, in his way, that he cared. I'd like to think
that.

We got up to a lot of mischief at the end of the
high school break after Ron left. Frank found a pair
of walkie talkies left behind by one of his brothers
before he'd moved out of home. We had a ball with
them all day and practically all night for weeks at

the start of the school break. We attached them to our bikes and road as far apart as we could while staying in contact with each other. Each day we increased the distance between us.

On one particular Sunday, on the hunt for new batteries for our walkie talkies, we headed to the local corner shops about two blocks away. Our route took us underneath an overpass bridge in front of Hebersham Public School. As we approached, we noticed a clear plastic bag filled with something lodged in a cavity underneath the west end of the bridge. Without thinking, we raced over and removed the bag. It was surprisingly heavy, like a ten-kilogram bag of potatoes, but it looked like it was full of charcoal.

We looked at each other for a minute or two and deliberated over the idea of a bomb fire. After a quick discussion, we decided to take it to my place for closer inspection. We needed to know what this stuff was.

By the time we got to my place, our minds were made up that it was charcoal, and we were definitely going to make a bomb fire. Leaving Frank outside holding the bag, I ran inside to get a box of matches from Mum.

"Mum, where's the box of matches?" I shouted out without thinking from the kitchen.

'Get out, no matches for you here and don't ask

me again now get out!' Mum yelled out loud and stormed into the kitchen before showing me the door. I was never going to get a box of matches from her and never to ask her again.

I returned empty-handed to find Frank already walking down the street, having heard my mum yelling at me. I quickly caught up to him and helped him carry the bag to his place. Once there, it was my turn to wait out the front while he went in to retrieve a box of matches.

Learning from my experience, Frank initiated a stealth operation by getting in the house using the back door to avoid any likely confrontation if anyone is there. His efforts were successful, so now we needed a suitable location to light the stuff. We thought of the new area in an adjacent suburb which was getting cleared out for new houses and headed there.

About two blocks away from our destination, we came across two younger boys about seven years old who asked us what we had in the bag.

"Why are you following, keep away." We both said to them directly.

We ignored them and walked past as if they simply weren't there. But when we got to a suitable cleared location in the new estate, we noticed they were following us now from a distance.

'Hey, keep away or else!' I yelled before placing

the bag down, turned towards the boys, and noticed they were still there.

It actually stopped them in their tracks and they looked on from afar. We didn't want little kids hanging around. We were older and knew what we were doing.

While I kept an eye on the kids, Frank opened the bag, grabbed a handful of the stuff, and placed it on the ground. We decided to light up a small amount to determine exactly what the substance was. Before doing so we both stood up and looked around carefully. Besides the two boys, there was no one else in sight.

We crouched down and Frank passed me the bag while he got a match out to light the small portion on the ground. Again and again, the match fizzled out.

After about ten attempts, we stood up, feeling frustrated. I then noticed that the two young kids were walking towards us. At the same time, Frank crouched back down and lit another match which he put into the corner of the plastic bag, just to see what would happen. Ignoring the two kids who were now about ten metres away. I crouched back down to see why this stupid material wouldn't light up.

A sound I'll never forget was accompanied by a shock wave that lifted us and threw us more than ten metres backwards. I must have been knocked out for a moment or two by the impact, and when I came to, I realised from the stench of burning flesh that I was on fire. I rolled around frantically, trying to put out the flames. After a couple of half a dozen rolls back and forth, I sat up to try and make some sense of what was happening and find out how badly I was burnt.

That's when the horror of this catastrophic event hit me. Frank, totally engulfed in flames, was lying on his back a few metres away not moving at all.

Flames suddenly flared back up on me and I quickly decided to take off my burning clothes. Until then, my hands were free from burns, but by the time I'd removed my shoes, trousers, and T-shirt they were severely burnt. My flaming T-shirt was the hardest to take off because there was a plastic picture of a car on the front which had melted onto my chest. I don't know where I got the strength to do that because the burns were so horrific, and I passed out a few times due to the pain.

I yelled out to Frank to do the same, but my words fell on deaf ears. The horror of it all had put him into a deep state of shock and he just lay there on his back, screaming for help. His agonised screams were unforgettable. By this time, I was sitting up, trying hard to stay conscious, while again attempting to peel the plastic off my chest with burnt hands. Then the horror worsened. The two curious little boys had also been caught up in the blast. They must have been knocked down by the explosion, as I saw them both get up with dark scorches on their clothing. Then, when I saw them running away, both engulfed in flames, I passed out, overcome with grief and despair.

I woke up in the back of an ambulance, lying in my undies with a burning sensation around the area of my family jewels. I tried to feel what was hurting me, but my hands were too burnt to be of any help. The ambulance, with lights on and sirens going, started up and headed to the hospital. On the way, the painful burning sensation in my crotch became unbearable. But my attempt to remove whatever was hurting me in that area failed miserably. I was burnt from head to toe but all that was bothering me was this excruciating pain down there until I lost consciousness again.

The next thing I knew, I was in the operating theatre at Penrith Hospital without any clothes on. To my delight, the pain around my family jewels was gone and I asked a nurse what it was that had been hurting me down there. The nurse replied, 'It was a small section of material still burning that was stuck to your undies which was slowly burning you and it is slightly blistered there.' Fortunately, she added, 'Everything is still attached, and all is okay down there.'

That relieved me.

The nurses and doctors rushed around frantically, sedating me, and making me feel as comfortable as possible. The doctors told me I had third-degree burns over seventy percent of my body. I was embarrassed, as all I had been concerned about was a little blistering on my family jewels.

As I was lying there, now almost pain-free, a doctor approached me and, in a very authoritative voice asked, 'Are you on any medications or do you have any illnesses that you know of?'

'Yes, I have thalassaemia major!' I replied confidently.

After a moment of dead silence, the doctor asked, 'Thalassaemia major, are you sure?'

'Yes, I receive monthly blood transfusions of 'A-negative blood!'

The doctor repeated everything I said and took down notes. Once he had finished, the nurses put a stainless-steel cage over me, which kept the bed sheet from sticking to me. By this stage I was pain free.

Before the doctors had a chance to take a closer look at me, I noticed Mum staggering with a look of disbelief on her face. I could hardly see her because my top and bottom eyelids were starting to stick together due to the swelling on my face.

Mum slowly approached me as I lay in the middle of the theatre. In a voice deep with anguish, she whispered, 'What have you done to yourself, Con?'

'It's me, Mum—Arthur. Don't worry, Mum. I feel fine,' I said, trying to soften her shock.

'Arthur? Is that you, Arthur?' she cried in horror and crumpled to the floor of the operating theatre.

The shock of seeing her child so disfigured due to swelling from such extensive burns was bad enough. But discovering it was me, who she had been trying so hard to keep alive all these years, was too much for her to handle. After she regained consciousness, she was quickly sedated for her safety and wellbeing, I was told.

As I took some comfort knowing they were looking after Mum, I remembered Frank. 'Where's my mate, Frank?' I screamed out loud in panic.

'Hello Arthur, I'm one of the doctors here in emergency at Nepean Hospital. Your friend Frank is in the same shape as you and his condition is also stable like yours. We have done as much as we can do for you both here and we are soon transferring you both by ambulance to the Burns Unit at the Prince of Wales Children's Hospital at Randwick. This is because they could treat your burns better there okay' I didn't know what to think of this, but I took some comfort in knowing Frank was alive.

But a chill ran up and down my spine when I recalled the other two young boys who had got caught up in the horrific incident. I was now fading in and out of consciousness and with a tremble in my voice, I asked the doctor, 'What happened to those other two young boys, Doc? Did they get burnt bad? I saw them get up and run off, you know. What happened?'

The doctors told me the boys had only sustained minor burns and were in for observation. However, they wouldn't be able to wear the same clothes when they went home.

The next thing I knew, I was travelling in an ambulance again, but it was dark outside this time. Apparently, our transfer coincided with a driver shift change, so, on the way to Randwick Hospital, the ambulance carrying Frank and I stopped on the side of the road and changed drivers. This didn't bother me, but I was fascinated with the sirens they turned on when we crossed through a busy intersection, which put me right back to sleep.

When I next woke, I was in a bed in intensive care feeling fine, without any sensation whatsoever. I was lying there without a stitch on and no sheet over me but covered with white cream from head to toe. Mum and Dad were there, both teary-eyed and completely exhausted.

After lying there for a while, speechless, I asked them to get me a mirror, which left them in a very tight spot. They didn't want to do it. I forced the issue, and they finally called a nurse to get a mirror. Knowing that what I was about to see would horrify me, Mum and Dad started to cry as the nurse held the mirror in place.

My whole face was dark grey and blown up like a balloon ready to pop. Both eyelids had been cut, separated, and taped back because they had stuck together earlier, due to the overwhelming swelling on my face. My eyebrows were gone, and my hair had melted onto my head in a weird way like melted plastic. My fingers felt like they were all tight next to each other but they had been surgically spread apart and were separated by wire. There was no skin on my hands due to me tearing off my burning clothes and my left arm had second-degree burns all the way up to my elbow. The plastic print on my T-shirt had left a third-degree burn on my chest and second-degree burn spots were scattered all over my stomach.

But what was most devastating for me to see were my legs. They were both covered in third-degree burns from ankle to knee.

Burns are classified according to the depth and

extent of the skin damage. With first-degree burns, the skin is red, painful, and sensitive to touch. The damaged skin may be slightly moist from leakage of the fluid in the deeper layers of the skin. For second-degree burns, the damage is deeper, and blisters usually appear on the skin. The skin is still painful and sensitive.

For third-degree burns, the tissues in all layers of the skin are dead. Usually, there are no blisters, and the burned surface may look anything from white to black (charred) or bright red from blood in the bottom of the wound. Because the nerves in the skin are also damaged, third-degree burns can be painless. When touched, the burned skin lacks sensation. It is important not to confuse a pale third-degree burn for normal skin as the lack of sensation or blanching of the skin blood vessels with pressure indicates damaged skin. A skin graft is usually necessary for significant areas of third-degree burns like mine.

When skin is burned, it loses its ability to do its job and protect the body, which increases the risk of infection. So the damaged area must be thoroughly cleansed, especially within the first six hours, and that the area is kept clean while it's healing.

I was lucky. After a few days, there was no sign of infection or the skin would have become increasingly red, hot, and swollen. But when I experienced throbbing pains, I feared the possibility of infection. Also, my severe burns caused a horrific amount of scarring and I feared I could perhaps lose both my legs.

In cases of extensive severe burns like mine, the body can lose large quantities of fluid. This can disturb the blood circulation and cause problems with the body's salt balance, so I was on a constant IV drip with an assortment of salt-based fluids and antibiotics around the clock.

After surveying the extent of my burns, I lay there naked in disbelief and asked the nurse to put the mirror away. All the while, I held back my emotions in front of Mum and Dad. I was bewildered as to how we had got to this point without killing ourselves. Looking back at the chain of events now, I can't believe we didn't question what we were doing. We just got caught up in the moment.

If we had only stopped and asked ourselves what we were doing or had shown a parent, then maybe we could have avoided this tragedy. As all young teens do, we got carried away without thinking, too headstrong and too impatient to consider the repercussions of our actions. The consequences, this time, were almost fatal for both of us.

I wondered why the small portion hadn't lit up after about ten attempts and why the bag had blown up after Frank had lit up the corner? These questions were going over and over in my head as Mum and Dad sat beside me in despair.

My body was pretty badly burnt, I was dehydrating fast and rapidly fading in front of my parents, but my mind was untouched. My thoughts were going a million miles an hour. I was determined to overcome this and stay alive because even though my body was wrecked, I knew in my mind I had what it took to stay alive.

———

The doctor's early successful attempts to clean and debride (remove dead skin and tissue from the burned area) my burns were outstanding. This procedure was done while I was in intensive care in a special bathtub using Lux soap flakes in hot water. They gently lowered me into the bathtub using a pulley (hoist) system.

I was overwhelmed at the thought of them putting me in a huge stainless-steel bath filled with washing powder and hot water to take off my blood-soaked bandages. The acoustics were almost deafening in that bathroom and, after I had let loose a few screams, it sounded more like a torture chamber.

This led to more IV fluids containing electrolytes, antibiotics by IV and by mouth, antibiotic ointments and creams in a warm, humid environment. Nutritional supplements and a high-protein diet were mashed up for easy consumption and force-fed through a tube. The pain medication was regimental and skin grafts were required to close the wounded area for functional and cosmetic reconstruction.

Skin grafts were taken from the top of both of my legs on the thigh area and grafted onto the damaged area, requiring fifty stitches on each leg. A skin graft is a piece of unburned skin, thick or thin, which is surgically removed from a non-burned area of the body to cover a burned area. Skin grafts are performed in the operating room. The burn that is covered with a skin graft is called a graft site.

———

Years later, we found out that a local industrial restricted site had been broken into and several ten-kilogram high-explosive bags of gunpowder had been stolen. Many of the bags were never recovered and investigators found that other highly classified explosives were also missing.

14

INTENSIVE CARE, 1976

After a couple of weeks in intensive care, I finally had a chance to see Frank for the first time since the accident. He was in the adjacent intensive care room with only a glass wall separating us. When it came time for us to see each other, they turned both our beds towards the glass and drew back the curtains.

With only that wall separating us, I had a clear view of him lying down, mirroring my position. I could hardly believe what I saw. It wasn't my friend I was looking at. This person looked horrific. I just lay there staring, trying hard to recall what Frank looked like, to recognise something that bore a resemblance to him. But the harder I tried, the harder it was to find any similarities between this person and Frank.

Then surprisingly, this person lifted his left hand slightly and started waving his whole arm up and down at me. With a lump in my throat and tears rolling down my face, I knew then that it was Frank, and I waved back.

We lay there for almost an hour looking at each other, bandaged from head to toe and each hooked up to an IV. We could only use our arms to communicate. It wasn't much, because of the fear of infection, but it went some way to filling up the emptiness I had been experiencing.

———

Excruciating, dangerous weeks turned into months, and we finally got out of intensive care alive. They placed us in a ward next to each other, both of us feeling sore from our recent skin grafts. Frank noticed I was receiving bags of blood via my IV drip as well as the saline solutions and other medications that we were both given. He also noticed I received a lot more doctor visits and that they paid more attention to me, but he knew why and didn't think twice about it.

He did get a bit annoyed that I got more visits

from family and friends than he did but he wasn't too upset as he would eat all my gifts after my visitors left. I didn't mind him having them after seeing that it gave him so much enjoyment. I guess he found his escape in food, whereas I had my mind focused on recovery and getting back to school. That's all I thought about and my mind was rushing a million miles an hour, thinking my way back to health.

I had big plans of going to university or getting into the Australian Air Force as a pilot or something like that. I was determined to make it despite what the doctors and everyone else said.

Funnily enough, Frank managed to get a kiss and a cuddle from Lilly, one of the morning-shift nurses. I couldn't believe it, even when I saw it for myself the first day they put us together. When Lilly started giving me the same kiss and cuddle in the morning, just as she started her shift, Frank wasn't impressed but soon didn't mind sharing. We both occupied ourselves with things that took our minds off our reality because it was too painful to deal with at the time and, after all, we were going through puberty as well.

Our skin grafts were not yet healed and even though they were bandaged and secured with long elastic socks, our legs were still not mended enough to walk on for more than a couple of minutes at a time.

'Arthur, did you know that your face is the fastest healing part of the human body!' The doctor nearest to me on one of the group morning visits leaned over and almost whispered it to me like it was a secret.

It wasn't until then I noticed both our first and second-degree burns were healing nicely but our faces were healing extremely well.

For two boys just hitting puberty, that was great news. Our legs had taken the full force of the blast and sustained the worst of our third-degree burns. Frank had third-degree burns from the top of his thighs down to below his knees and I had them from above the back of my knees down both legs to the ankles. Poor Frank, on the other hand, didn't have as much unburnt skin to work with, so the doctors had to use skin grafts from his bottom and graft it onto his legs.

Frank and I owned that section of the burns ward. It was our domain. When it was time to start physiotherapy, the doctors put us in wheelchairs. We were now mobile! The doctors planned to give us some time in wheelchairs to help us regain our middle ear balance after lying down for so long. The aim of the customised wheelchairs was to make us mobile without bursting stitches or blood vessels. The nurses placed a stiff sheet of hardboard on the seat of each wheelchair to immobilise our legs, which were stretched out straight in front of us, parallel to the floor.

Giving us wheelchairs was a mistake, let me tell you! We were unstoppable in them. There was no end to the mischief we got up to and the chaos we caused. These came in useful as battering rams when we challenged each other to a duel to the death.

Physiotherapy and daily baths brought us painfully back to reality but as soon as they were over each morning, off we would go on another adventure in our wheelchairs. It would take our minds off the horrific situation in which we found ourselves. We didn't have time to get upset as we kept ourselves busy with whatever we could get our hands on. If I'd stopped for just a minute and thought about the risks of dying from complications of an infection because of an unhealthy immune system due to having thalassaemia major, I would have buckled under the pressure and sunk into the dark abyss of despair.

It soon came time to walk and, even though I had been doing it for practically all my life, I didn't have a clue how to keep myself from falling over. At that stage, Frank's grafts weren't healing well, and the burns doctors decided to harvest new skin grafts from his bottom. By the time Frank had healed enough to attempt to walk again, I no longer needed any assistance and was able to help him to walk.

My recovery was progressing well and when I got the okay to go home, I was thrilled. I was disfigured, battered, and bruised with a slight limp but I was going home. Frank needed to stay in for a few more weeks, though, which was a setback. Even though we tried to make it like an adventure, especially when we'd got our wheelchairs, the experience as a whole was horrific and terrifying.

We had been admitted to hospital in the January school holidays and it was now the middle of the year. I couldn't believe it was June. Was I in a time capsule or was it a nightmare? It was both, I guess, but I concentrated on looking forward, which made everything a lot easier to deal with.

The doctors explained that I must keep on wearing the elastic bandages to avoid bursting blood vessels or detaching the skin grafts. Furthermore, for the next two years, I wasn't allowed to expose any part of my burns or the sensitive graft areas to sunlight or UV rays.

There goes swimming and running out the window. But I was grateful to be alive, so I persevered without any regret and, with a solid embrace, wished Frank a speedy recovery. I then said goodbye to all the wonderful nurses and doctors, including all the Thal specialists and other doctors for their efforts in handling my thalassaemia treatment in the background while all of this had been happening. This wasn't my hospital and yet they'd treated me as if I'd been going there for years.

I collected my things and headed home with both my parents. On the car trip home, cheerful conversations were interspersed with long silent periods. Suddenly, I felt ashamed for putting my parents through all this trauma and, for the first time in my life, was not sure of myself. I felt as if I were coming home burnt, battered, and completely humiliated from a war I had started.

At that moment, I knew I needed to learn from this horrific mistake if I were to survive. I identified that my confidence was really arrogance, and my sudden feeling of shame was not to be ignored because it was telling me to learn from this horrifying experience. I also knew I shouldn't dwell on it but had to hold it close to me to remind me in future to question myself in every situation to ensure I was doing the right thing.

We finally arrived home to silent stares and a smattering of sympathy from my brothers, which only lasted a brief minute or two, followed by a bombardment of questions about my burns. After

their curiosity was satisfied, they went back to ignoring me as usual. It was great!

It was hard to get around in those early days at
home after Mum and Dad went back to work. The
hardest thing by far next to walking was the tor-
turous process of removing the blood-soaked elastic
bandages in the bath, which would unintentionally
cause more blood blisters to burst. By the time
Frank was released from hospital a month later, I
was efficient in applying and removing the elastic
bandages and I was able to walk unassisted to
Frank's place around the block.

Frank was thrilled to see me the day he finally
came home, almost as thrilled as I was to see him.
He was flabbergasted to discover I had walked to his
house unassisted, which inspired him a lot. His recovery at home was more rapid than my sad state of
mending.

Even though his healing progressed well, I noticed something different about his behaviour that I
couldn't quite put my finger on. Anyway, no matter
what it was, I was glad to have him back home.

After about two weeks, Frank caught up to me
and started walking without any assistance to my
place as well. We were both well on our way to nor-
mality once more and we felt good about it. It was
great spending the days at his house and my house

while our parents were at work and our siblings were at school.

———

Frank's parents had separated well before we'd got burnt. They divorced without delay and Frank's mother established a lifelong partnership with a huge Aussie called Barry. Just as Frank's mum had always done, Barry quickly welcomed me as part of the family. Both Frank and I were already taller than Frank's mum, but Barry was a giant of a man who was well over six feet tall and weighed almost as much as a small car. He was enormous, with a dreadful temper to boot. If anyone dared to piss him off, the whole house rocked back and forth on its very foundation, but otherwise, he was a true gentle giant.

Most of Frank's siblings had left home by then, so Frank's house was all ours to hang out in for the whole day. Our days were filled with midday movies, trying to perfect a good cup of coffee without ten spoonfuls of sugar and eating toast to our hearts' content. We never got bored with toast because we experimented with any type of food that was consumable as a topping and washed down our culinary creations with numerous cups of sweet coffee.

'Frank, check this out.' I would often call out to him from the kitchen hoping to surprise him with a new flavour of coffee. But he knew me well and always kept his distance. That didn't stop me from continuing to perfect the ideal coffee at every opportunity I got.

15

BACK TO HIGH SCHOOL, 1977

The days we spent healing at Frank's place eventually came to an end as our burns healed to a point that it was now time for us to go back to school.

The trauma of our morning routine of hot baths to remove dried-up blood-soaked bandages began to lessen. We both became more efficient in changing our bandages and elastic dressings once our wounds had healed enough to stop bleeding. Yes, we both were still tender from our other minor facial and body burns but we knew it was time to go back to school and didn't make a fuss.

Frank was self-conscious about the coin-shaped piece of hair missing from just above his hairline in the middle of his forehead. Given that his hair, like mine, was still short after having been singed right off, it was hard to hide what looked like a hole in the front of his head.

I, on the other hand, was self-conscious about everything from my burnt hair and red raw vulnerable skin, through to my critically damaged legs. This was a strange period in my life when, for the first time since finding out I needed to receive regular blood transfusions to stay alive, my Thal took second place.

The day finally came to go back to high school. Our ruined bodies were mending better than expected. We started in Year Nine in 1977, not knowing what to expect after missing the entire previous year of school.

Our legs were healing at a tremendous rate, and we were both able to walk to school. We still didn't look the best, but with our bandages, elastic dressings and UV protective sunblock cream slapped on all over, we were ready to venture back into the schoolyard. We braced ourselves for the worst.

'Hey, Frank, my legs are itchy again.' I kept repeating as I was trying to keep up with his stride.

'I know, me too but you need to leave it alone. You know what the doctors says.' Frank said without turning to face me now picking up his pace.

'Why are you walking fast?' I called out to him.

'I'm not, you are just slowing right down. Come on, we are going to be late.' Frank answered back that got me to finally focus now on walking instead on my awful itchy legs.

We managed to walk to school slowly, and the day started as it ended, quietly and free of any mishaps or teasing. We couldn't believe our first day back had turned out this way at our high school.

Neither of us had minded the stares we'd received from everyone or the weird questions, but we did mind being put in the lowest classes in our year. After blowing ourselves up, we may have looked like we were dim-witted, but we took offence at being put in the dumb class.

'Hey, we would be the smartest in the class and we would shine by scoring the highest in these classes, so let's tried it out for a while.' Frank explained.

We debated the positives and the negatives of this situation a bit further for a while. At first, I could see he was right and even though we were split up between the two dumbest classes, my classmates soon looked to me as a support person to help those who couldn't finish the tasks allocated by the teacher. I felt enormous pleasure and satisfaction in being able to complete any task given in a fraction of the time of all my classmates, then being asked to tutor the class while my teacher left early or was late. It was great.

About the third week back at school, I was lining up with the rest of my class outside the science room when the inevitable happened. A boy from my class pushed me out of the line and knocked me to the ground. 'Get to the back of the line, wog!'

I picked myself up, quickly brushed myself off and stood two inches from his nose. 'Do you want a fight, mate?'

'Yeah, I'll fight you, wog! You're dead, mate!' the boy replied, filled with resentment. Who knows what had set him off, but the 'wog' comment must have been about my surname because everyone thought I looked like an Aussie.

I didn't back down. 'Any time, any place, mate!'

'After school behind A Block, wog,' he said with a smirk.

As we stared each other down, the science teacher arrived, separated us, and calmed the rest of the class down. He then marched us all into the science lab. 'I will have none of that sort of thing going on in my classroom. And if I find out that you have taken this elsewhere, I will come after the ones who disobeyed my instructions with unparalleled consequences.'

The whole class went on with the lesson as if

nothing had happened, except for the silent death stares exchanged between my adversary and me across the classroom. This situation was unusual for me, and it was futile to attempt to write down or retain any of the information provided by the teacher in that lesson. All I could think about was the likelihood of me getting killed after school by this boy I didn't even know.

Frank had taken that week off school because of ongoing problems with his legs and stomach, so I was left all alone with no support or back up. The friends I'd had before the accident were all in the next grade in classrooms on the other side of the school and the end of the school day was fast approaching so I had no time to contact anyone. I was shit-scared at this stage but by the time the last bell rang for the day, I made my mind up to follow through and front up to the fight like a man.

Despite my body not being totally healed, especially my legs, I knew I would have a hell of a time being teased for the rest of my high school days if I didn't turn up to this stupid fight. So I took a deep breath, held my head high and proceeded alone to the fight location.

I couldn't believe my eyes. The whole school was there, all screaming for this boy to kick my arse. Spellbound, I wondered what on earth I had done to make the whole school hate me. With my knees almost buckling, I kept a steady grip on my nerves and continued towards the angry crowd. Feeling as if I was heading to my execution, I became inexplicably angry and started to walk faster and faster, showing no fear whatsoever.

When I reached the edge of the rowdy mob of students, the crowd parted, leaving just enough space for me to walk through. In the centre of the noisy crowd, the boy who had started all of this stood waiting for me with both fists raised. Without blinking an eye, I dropped my bag and strode right up to his face, frightening not only him but also myself with my intensity.

In response to my actions, the crowd roared. Then I had a stroke of luck. Because I was so close to him, I noticed the fear in his eyes. The poor boy was shocked by my abrupt approach. Obviously, he had not expected this from me. With adrenalin pumping through my veins, I could feel no pain, so I pushed this guy and said, 'You picked the fight, so you throw the first punch!'

Again, the crowd roared and screamed, 'Kill him!' to me, which frightened this guy more. He stood over me, ready to take me out with one punch, probably thinking that was all it would take. But he hesitated. So instead, I struck first and took him out with one lucky punch. A king hit to the face laid him flat on his back. The crowd stood there in silence for a second or two, mesmerised by what had just happened.

Thrilled and scared at the same time, I stood there waiting for some reaction. Completely taken aback at how lucky I had been, I quickly re-focused when the crowd started to murmur. My opponent got back up on his feet, bloody and more hesitant. But then with no warning at all, he grabbed my legs and tackled me to the ground. The pain was unbearable, and I screamed for him to get off my legs. He then jumped onto my chest to try and hold my arms down. That's when I held his head down onto my chest with one arm and repeatedly hit him in the face with my other free hand. I took my agony out on his face as his blood splattered all over me.

After I connected a few good hits, a teacher came from nowhere and grabbed the boy by the back of his neck and lifted him off me. The teacher thought it was my blood because the other boy was on top of me and he was taken away to the principal's office, kicking and screaming. The crowd immediately dispersed, and I was left there alone lying in a cloud of dust.

The legend status from my unexpected king-hit was quickly forgotten, replaced with the comprehension I was lucky to still be alive. I picked myself up and dusted myself off and that's when I noticed my bandages and elastic dressings were completely soaked in blood. The throbbing pain in my legs was unbearable and I didn't know how I was going to walk home. But I had no other option, so with all the strength I could muster, I picked up my bag and trudged home, one painful step at a time.

By the time I got home I was so weak from the pain of my throbbing legs, I collapsed onto the bathroom floor. I lay there for almost an hour before I regained enough strength to crawl into a hot bath. As I started taking off the blood-soaked bandages, I was horrified to discover my skin literally peeling off, so I panicked and immediately put the bandages back on. As I lay back in the bathtub, I noticed the water in which I was soaking had turned red, which panicked me further. I couldn't believe what was happening and tried to calm myself down.

No one else was home, so I needed to get a grip on what was happening, or I would be in serious trouble. I realised I was bleeding severely, and I needed to slow my heart rate down and elevate my legs to reduce blood loss. I had no time to spare. I didn't want to turn my bloody bath into anything more serious, so I composed myself and relaxed and rested my ankles on opposite corners of the top of the bathtub to elevate my legs.

Another half an hour passed before my blood-soaked bandages came off themselves, thankfully leaving my new skin still attached to my legs. The wait had paid off. I reached over to the bathroom cabinet to retrieve new bandages and wrapped my legs, taking care to cover up the areas of broken skin. After covering my new bandages with new elastic dressings, I had just enough energy left to clean up the bloody bathroom before anyone else got home.

16

WORKING AND DISCOS, 1979

Graduating high school was not a highlight for me. I was glad to get the hell out of there in late 1978. I was looking forward to working and going to discos.

I got a job as a real estate agent. I quickly got my selling licence, but my boss wanted me to get a car of my own before starting my real estate qualification and going full-time. After working six weekends in a row, I managed to save a little over a thousand dollars for the car.

Confident my dad would find me a bargain, I handed all my savings over to him and he didn't disappoint. I became the proud owner of a yellow VH 1972 Chrysler Valiant four-door sedan. It wasn't perfect—far from it—but it was mine. I didn't care that Dad was telling everyone he bought it for me. I had a car of my own and I loved it.

Once I had my own car, my confidence soared. I was on my way to getting my first full-time job. But while my boss was good at dangling a carrot, he didn't follow through. In fact, he soon replaced me with another sucker who'd just left school, promising him the world. The weekend work had been good but by then I wanted more, and I was sick of all the boss's dodgy excuses for paying me less than half the agreed commission on my sales.

Finding myself unemployed wasn't an issue because I had a steady income—I had been on the pension since leaving high school. Back then, it was called the invalid pension. Nowadays, it's called the disability support pension and my dad's proudest day was when he was successful in getting my application for the pension approved. His ambition for me was to get permanent government financial assistance.

'See what I do for you, you haven't a care now so enjoy your life!' Dad said very proud with himself. Mum gave him a hug and a kiss before they headed for the kitchen so she could sever him up his prize for a job well done.

'What did you do for me?' I asked not surprised he was up to something.

'Your father got the doctors to finally sign your

pension application, so how about thanking him. Now you never need to worry about getting a job for the rest of your life.' Mum announced it all in the kitchen while putting on her apron.

'Why didn't you tell me first?' My reply fell on deaf ears.

It wasn't easy for him to do that, and he did it without consulting me whatsoever. The more my doctors resisted his requests to support the application, the more he pushed. Despite not being consulted about my financial future, after seeing the glow on his face on the day he first told me I was going to receive financial assistance for life, I was happy for him. But the penny didn't drop for me at the time that this was at the cost of me never being able to work full-time for the rest of my life.

When I confronted him with my concerns of not ever being able to have a full-time job, he reminded me how horrible it had been arguing with all the doctors to sign the forms for me to get on the pension. He was genuinely happy for me, comforted by the thought that I would be looked after financially for the rest of my life. But he had locked me into a life of poverty and permanent frustration. How could I explain that to him when I saw how happy he was with his achievement?

With no prospect of ever working full-time, I started hitting the discos in my Valiant with Frank.

He was working full-time cutting fabric for garments in a factory, earning so much money he bought an 18-foot boat, in addition to an almost new, four-door Chrysler Galant. He was also saving to buy a house.

But every Friday and Saturday night, the Zoo disco nightclub on William Street near Kings Cross was our regular destination. That and two other hotspots in and around Sydney. But the Zoo remained for us the weirdest and most exciting place by far. The girls and entertainers were astonishingly dancing almost naked with only what looked like a G-string on, and it was also the place we first met Max Markson, PR man extraordinaire.

'Hello, my name is Max. Where's your roller-skates? It's roller disco tonight!' Max screamed at us as he rolled by and around us just as we entered the nightclub.

'But it's Saturday night, isn't it on Friday nights?' I replied screaming back totally amazed with his roller-skating techniques. We all couldn't stop watching him skate around the entire nightclub.

'That's right, so, here's some free tickets for you wonderful people. Remember, it's on next Friday night, so see you all there!' He announced after circling around us for a while before he was off again.

Max was only a marketing promoter back then and he often gave us free tickets to the Zoo disco whenever he saw us. Later in life when I tracked him down to seek his help in launching my first self-published novel, he couldn't recall meeting me. Then again, it was years later and by the time I caught up with him he specialised in promoting big-name celebrities, so it wasn't surprising he couldn't remember me.

It was outside the Zoo one night that Frank and I narrowly escaped death. After an awesome night, we encountered two massive guys. The bigger one punched Frank and he was extremely lucky the king hit didn't fully connect. It was a drunken cowardly random attack that we didn't see coming until it was too late. Shocked by the force of the blow, Frank managed to avoid serious damage or even death by anticipating the punch and moving away at the last second.

Bearing only a grazed bruised cheek, he took up a Bruce Lee stance.

I jumped out in front of him. 'Don't do it! You know what our Kung Fu sensei said, your hands are lethal. You will go to jail if you kill someone!'

The giant hesitated. 'Look, man, I don't want trouble, okay? Sorry, I thought you were someone else.'

'He's full of shit, man,' his mate said, thirsting

for blood. 'Fuckin' kill them. You can take them both!'

'Look, you fuckin' idiots, you're lucky I can stop him at this point, okay? Fuckin' leave now or I'm letting him go. It's your call!' I screamed as loud as I could.

I was hoping to get the attention of the Zoo security guards who were just out of sight but, to my surprise, this colossal guy broke out of his boxing stance and walked away, dragging his mate with him. We bolted off in the opposite direction and didn't look back.

Although on a high, we were so paralysed with fear that we got to the car and drove away in silence. In fact, we didn't talk about how our spontaneous coordinated role play had saved our lives that night. Neither of us knew Kung Fu, but we loved Bruce Lee movies and we have them to thank for our lives.

Despite our brush with death, we loved discos and were back again the following week. However, from that night on, we ensured we always left with a crowd.

17

GIRLFRIEND, 1983

At high school, I hadn't had any girlfriends. I guess their radars were never tuned in my direction. The closest I ever got to a girl was in my first year of high school. It was a harrowing experience that kept me away from the opposite sex for quite a while.

I had been heading out of the classroom into the playground for lunch when a girl from my class came out of nowhere and started pulling my hair. She didn't let go until she'd attracted an audience who finally told her to let me go. I didn't know what was more upsetting, the spin she put me in or the handful of hair she'd removed. *If that's what you get just walking near a girl,* I thought, *I'd hate to imagine what I'd get if I ever asked one out on a date.*

It wasn't until I'd left high school that I started dating girls, but nothing ever came out of any of these dates. I guess things didn't work out because I was constantly worrying about one thing or another, mainly about my condition. Then, early in 1983, just before my twenty-third birthday, I was introduced to Helen Didaskalou by a young couple I knew, Vlado and Dawn. Helen happened to be friends with Dawn, and I was friends with Vlado who only live down the street from me.

It was at Penrith Leagues Club early one Saturday night and I was on my way into the venue's disco when I first saw Helen walking in with Vlado and Dawn. Helen was twenty-one and gorgeous, but before I had a chance to get a word out, she looked me up and down and said, 'Not another wog!' She walked away with a grin.

'This was Helen's first time out with them, her strict parents hardly ever let her out at all.' Vlado and his fiancée Dawn explained collectively and finally got her to go out with them.

I was trying to listen to them but was hypnotised by her beauty. Awe-struck, I ignored her comments after they introduced me and followed them in. I couldn't keep my eyes off her.

Once inside, I was grabbed by my other friends who were there waiting for me, and we raced on to the dance floor just as a Bee Gees song started playing. The floor on this particular night was packed to the edges but I still didn't lose sight of Helen through the crowd. I took control of the floor, which shocked the girls I was dancing with, but it was the only way I could keep an eye on her. I didn't care so I went for it. I knew Helen couldn't see me from where she was sitting, so I tore up the floor, keeping one eye on her always.

Once the song had finished, I went to find a seat near Helen's table, but as soon as I got there she was on the dance floor herself and she danced the rest of the night. I was devastated that I didn't get a chance to talk to her before she left later that evening, but I asked my friends who were with her to let me know the next time she went out with them.

Three months later, they finally told me she would be going out with them again. Filled with excitement, I couldn't wait to see her. My heart was pounding. It felt like a lifetime since I had seen her, and I needed to confirm she was real and not someone I had imagined.

When the time came to head off to the disco, I found it hard to leave the house as we had visitors. My cousin Effie and her husband Steve Rallios, her mum (Aunty Dina) and dad (Uncle Spiro) and her brother (Cousin) Andrew Bozikas and her sister (Cousin) Angela Bozikas were visiting from Melbourne. Desperate to leave, I slipped out the back door at the first opportunity.

'Where do you think you're going, cuz?' Effie grabbed me from behind before I got to the car.

'I can't talk, I need to go see about a girl and I'm running late.' I replied in a hurry.

Effie was a few years older, I had been a groomsman at her wedding five years earlier. I couldn't get away until I spilled my guts to her. Once I'd told her about Helen, she jumped in the car, informing Steve and my Uncle and Aunty that she wouldn't be long, insisting I show her who I was talking about. She loved discos too and wanted to go dancing. I didn't mind—I just wanted to see Helen. By the time I told Effie all about the night I'd first seen her, we were there.

It wasn't too long until I spotted Helen dancing in the middle of the floor with my friends. A sense of calmness come over me. She was real and not just a figment of my imagination! She looked so beautiful under the array of colourful lights. The more I stared, the more my heart filled with joy at seeing her again.

'So she's the one you're going to marry, right?' Effie screamed for all to hear.

I didn't care how loud Effie was, I just nodded.

I soon left to take Effie home and on the way, she was forced to endure my lovesick ramblings. I told her of my hopes, dreams, and expectations. She was a real trooper and was genuinely happy to listen to me. I also loved how she encouraged me without reservation because I knew Helen was the one.

A few weeks later, I heard there was a strong possibility my friends would be taking Helen back to the same disco the following Saturday night. This was music to my ears! In my excitement, I told little Arthur about Helen and he agreed to support me to ensure I could talk to her this time. Maybe having him there as a backup would help me to find out if I had any possibility of a future date.

When the night finally arrived, confronted with her beauty face-to-face, I found myself tongue-tied. It was little Arthur who got her talking and asked her to sit at our table, which she surprisingly did without hesitation. I loved him for doing that, but I could have killed him for sitting between us.

'So Helen, do you have any sisters?' Arthur asked with much anticipation.

'Yes, one older than me and married.' Helen reply with a smile.

It was amazing watching them both talk so much. I just sat and watched her until they finally stopped talking so Arthur could get her a drink. That's when I took my opportunity and pounced on his now vacant seat. With some restraint at first, so not to overwhelm her, I started talking to her. After a couple of minutes, I took my chance. 'Have you seen the movie *First Blood* with Sylvester Stallone yet?' I asked casually.

'No, but I've heard all about it and wouldn't mind seeing it,' she replied wide-eyed with a grin.

'I haven't seen it yet either. It's still playing at the movies. Do you want to go? We can see it together if you like?' I said, staring into her beautiful brown eyes.

She agreed with a smile, and I arranged to meet her in front of Vlado's place the following week. I had managed to do all of this before Arthur returned with the drinks. Once he was back, I happily returned to my seat and was content to remain there for the rest of the night. It didn't matter that I didn't dance with her, have a drink with her or even have a lengthy conversation because I had arranged a date with Helen and that's all that mattered.

'Hey Arthur, Helen liked me more, sorry!' Little Arthur broadcast in total joy.

'Sorry cuz, but I have a date to the movies with her the following week.' I replied as we both headed out at the end of the night.

Little Arthur took it well but didn't feel like talking on the way home.

18

LITTLE ARTHUR, 1983

Early in 1983, Michael Jackson's *Thriller* album had only recently been released and for me, this music was life changing. Living in my room every night and all weekends, drifting apart from Frank, the music forced me back out to the discos just like the Bee Gees had in the seventies.

Now little Arthur was over eighteen and I had him for company on Friday and Saturday nights. Having him with me the night I asked Helen out on our first date was great because it gave me someone to earbash about Helen. I must have worn him out, talking about her every single time we met, but he was wonderful with his encouragement.

From as far back as I could remember, little Arthur's parents, Uncle Tony and Aunty Mary, had a takeaway business. I loved it because they would offer me whatever my heart desired. They welcomed me and treated me like royalty. Both my Uncle Tony and Aunty Mary, my mother's younger sister, always had something kind to say.

His true talent was making me laugh; it came easy for him. What I loved most about little Arthur was how he never considered me to be sick. All my life I have been referred to as 'the sick one' and my relatives were the worst. No matter what the occasion, they went out of their way to track me down while screaming out as loud as they could about my condition, often making a scene.

Any attempt to keep my condition secret from others was impossible due to my family, especially my mother but she earned it. Not only did she go out of her way to inform anyone and everyone about my condition, but she also made it her life's duty to tell people I wet the bed as well. What she failed to explain was that my bedwetting revolved around the times I was most weak and lethargic, just before my blood transfusions.

My mother's other sister, Aunty Helen and her husband Uncle Angelo, were in business longer than little Arthur's parents. Throughout my teens, I thought they were the richest people in the world. They had money but what made them rich was how they raised their children. I also loved how my

Aunty and Uncle didn't treat me any differently from their own children. Their oldest, my cousin Diana, was about ten years younger than me. My cousin Voula had followed a few years later and soon after that, cousin Leo. They are now all wonderful parents themselves, all married to beautiful partners raising their gorgeous children.

From a young age, my three cousins were like wise old owls. It was amazing watching them grow up oozing confidence, respect, and kindness beyond their years. I cherished that period, and the experience was a gift that I absorbed as much as I could and later used in raising my own family. Both my Aunties and Uncles have remained incredible people throughout my life.

However, I always gravitated towards little Arthur, as he made me laugh. It was insane when he was finally able to come to discos with me. The dance moves he came up with were legendary. At every opportunity, he would get up on the dance floor and people would gather around and be amazed. It was far from a John Travolta style—more like Elaine Benes from the TV show *Seinfeld*. He left everyone in stitches everywhere we went. On occasion, he took it up a level and brought the house down with his body rock routine on his stomach, where looked like a cross between a rapper and a wounded dolphin.

I loved him for getting me out of my room and out of the house. Although he was just a little over five feet tall and I towered over him, to me, he was six feet plus.

Jim, Little Arthur's younger brother by almost ten years, grew up to be just as spirited. I tried to hide him in the back of the car on our disco nights, but Arthur would always discover him at the last minute. Jim couldn't wait to come out discoing with us but it suited Arthur just fine not have his little brother trailing after us.

Sometimes I waited for ages for Arthur to finish his work at the shop and get ready. I always turned up early to watch TV with his mum. She never took any time out for herself but loved watching *I Love Lucy* on TV.

My Aunty Mary was always serious like a busy bee, so getting there early to watch her laugh was a treat. I was mesmerised watching her laugh at Lucy's antics. It didn't matter what episode was on, she would be in stitches. In fact, she started giggling even before she turned the TV on. It was only the two of us together, uninterrupted for thirty minutes, and I loved that she allowed me to do that with her.

Arthur's dad, on the other hand, reminded me of a flamingo— my Uncle Tony always took large steps and boasted an array of flamboyant colourful leather shoes. Watching my Uncle walk was mind-blowing and I could be spellbound for hours.

———

My parents, after working hard all their lives, decided to follow both my Uncle's and Auntie's path and purchase their first business when I was in my late teens, early twenties. Nick, with his McDonald's contacts, tried to turn our fish and chips takeaway into a fast-food business by implementing fancy packaging and slick uniforms. My younger brother Con, at nearly eighteen, and the baby of the family, Angelo, at thirteen, were not impressed either but we all made a go of it, determined that Lalor Park Takeaway would be a goldmine.

The idea was great, but my parents didn't put enough thought into it. From a young age, my brothers and I hadn't worked well together. Now we were older and further apart than ever, but we were forced to work with each other.

It was bad enough living together in the same house, making us work together too was a disaster in the making. The fights in front of the customers, although far from ideal, weren't the real issue—it was the mismanagement of working different shifts that really damaged the business.

Working seven days a week bearing black eyes

and bruises was bad enough, but noticing the sales decreasing regularly after implementing all the changes that Nick put in place was heartbreaking. It wasn't all that bad because the business was profitable, it was only our attitudes that failed ourselves and our parents. As I watched our dream of a goldmine slowly fading after selling the business, I saw the parallels with my life and was reminded of how lucky I was to still be around to be experiencing this bitter sweet episode.

Hitting the discos every week with little Arthur gave me a temporary escape from all my worries. The constant fighting with my brothers also helped me forget about my condition and what was in store for me. The fear of death only bothered me on those dark nights just before I fell asleep. However, the daily antics we all received from Nick, combined with the wild nights out with little Arthur, helped to minimise those dark nights and I loved them for it.

19

MIRACLE TREATMENT, 1983

Right after my twenty-first birthday, it was my turn to begin a miracle new treatment. I would be the last of our group of Thals to spend a week in hospital to be taught this new method to remove the excess iron from my body and thus extend my life.

The accumulation of iron from all the blood transfusions was the reason why Thals didn't live past their early twenties. The impact of iron overload had been clearly visible to everyone since we were young children, with our dark skin coupled with a yellow tinge in the whites of our eyes. Throughout our lives, we'd been told it was impossible to remove the excess iron from our bodies and we were all now at the last stage of life, waiting for the end. My surviving Thal mates and I were fast running out of time.

After first being told about this new miracle treatment, I was reluctant to agree to it because the specialists made clear to my parents and me that this treatment might not be effective at this late stage of my life as a result of damage already sustained to my organs.

'I'm sick being here, why I'm I still in the children's hospital?' I belched out in total frustration in front of the doctors and my parents.

'Look, you need to focus on deciding about your treatment just like your friends have. I have given the information to your parents and take your time getting back to me please.' The doctor explained before gathering back with his team before leaving me to talk it over with my parents.

Why bother? I thought.

Initially, I was furious and refused the new treatment. But Severino, George and Peter, the only ones of the older Thals left, opted to go ahead and I watched their complexions as they looked more and more normal every day. Although I wasn't looking forward to what was in store with the treatment itself, with their encouragement, I finally decided to give it a go.

This new miracle treatment was called iron chelation and it was a process used to remove excess iron from my body. Using Desferrioxamine (Desferal), I was introduced to a low dosage at first. I loved learning how to mix it up with vials of water for injection, but I didn't like injecting myself with it. After filling a 10cc/mL syringe and attaching a butterfly needle, I had to insert the needle subcutaneously, preferably around my stomach area. A small pump machine powered by a 9-volt battery then pumped the solution under my skin over a period of eight to twelve hours.

After setting it up and strapping the needle securely onto my abdomen, I placed the pump into its sling around my waist and forgot about it until it was finished. While in hospital that week learning about it, I became good at it and would put it on at all hours of the day. When I got home, I only used it at night.

I quickly made it my routine to set up my treatment one hour before bedtime. This worked well for me because it was all done by the time I got up the next morning and it freed me from the burden of doing it during the day. Desferal works because it binds the iron and then gets rid of it through the urine, thus removing excess iron from the organs. By the time we all started using Desferal, our bodies were riddled with iron, and it was oozing from our pores. We all looked astonishingly weird because we all look so dark it was frightening because it was an abnormal ashy colour like burned wood.

At first, my compliance in administering the treatment was better than average. But after a while, I started thinking about what my specialist had said, that it might not work anyway at this late stage, and I started slacking off. I had been given clear directions to do the treatment at least five nights a week. Six months into it, I was down to four nights. Twelve months later, I was lucky to use it twice a week.

Working in the shop seven days a week, plus hitting the discos with little Arthur on weekends, left me with little free time. In addition, I could hear my specialist's warning repeat over and over in my head that the damage was already done. *What's the use?* I thought, *I'm going to die anyway. It's bad enough I have to get regular blood transfusions so why put myself through this additional pain and discomfort every day?*

A side effect of the Desferal treatment is that the colour of the urine is transformed into a deep red colour. This shocked me when I saw it. I panicked, thinking I had internal bleeding. *This is it,* I thought, *my end has finally arrived.* It took me a while to remember it was the Desferal and that jolted my heart back into normal rhythm.

On occasions, George, Peter, and I met at a

local leagues club to have lunch and play a few rounds of pool. It wasn't unusual for Peter to wear his pump of Desferal during the day, even when joining us for lunch. Having had his pump running since before leaving home, he waited until after his lunch to visit the toilet; a peak time of the day for patrons to visit the men's room. After strategically placing himself at either end of the packed urinal and never in the middle, he began to urinate deep blood-red urine that trickled down the urinal in front of those poor unfortunate souls trying to relieve themselves. Whenever we heard screams and shouting from the toilets and saw a mad stampede of patrons rushing out, we knew Peter was at it again.

Quite soon after you stop the Desferal, the urine colour quickly fades back to normal. But in these instances, Peter's pump had been running for hours, resulting in the deepest red possible. Some unfortunate victims received such a shock they wet themselves trying to escape Peter's gruesome and apparently diseased urine.

Seldom did people ignore him and some even tried to help him, genuinely concerned for his welfare. Either way, it was a reaction, and that's all he was after. In the beginning, we found it funny, but he was constantly seeking attention any way he could, and this tested us all on many occasions.

Twelve months down the track, my skin was noticeably lighter than ever before. This improvement in such a short time perked me up and got me back on track with my treatments, but never to the recommended five days. My obstinate and stupid reasoning was overwhelming and interfered with my compliance.

Dr Bau was wonderful with us when anyone else would have given up on us.

'I forget to use the pump some weeks, but I'm back using it now.' Peter unashamedly would announce.

'I thought it was just me?' I would reply.

'Don't worry about it, you are back using it now so it's okay.' George would reassure me on many occasions.

Our compliance with our Desferal treatments was dreadful, but he came up with innovative ways of getting it into us safely on our blood days to avoid treatment that night. Although it was all done in hush-hush, but he somehow got our treatment into our blood packs on our blood days, genius!

One day he took us all to Chinatown in the city and shouted us lunch at his favourite Chinese restaurant. He was a treasure, not because of lunch, but because he'd treated us like adults from the first time we'd met him. He was a once-in-a-lifetime doctor who had come to us when we'd needed him the most.

It was via his efforts we all got on Desferal. As soon as he got approval for all of us, he got us all on the treatment straight away. And it was his calming manner that put us on the right track from time to time. I admit I couldn't look into his eyes at times and be honest about my compliance.

He was always patient and encouraging with everyone and some of that eventually rubbed off on us. We started talking to each other and, after a while, that grew into something consoling. At first it was only with one of us and in time it was great to see everyone talking about it all in the open. We were all maturing, and the jokes suddenly stopped, replaced with uplifting remedies and wise advice from each one of us. Before long, we noticed that George's advice was by far the best. Advice soon turned into caring and, before we knew it, we were all monitoring each other at every blood day.

20

THREE MUSKETEERS, 1983

By now we were adults and routinely receiving blood transfusions every four weeks, experiencing a better quality of life. But Mary's death continued to rattle all of us. I only had two beautiful years with her. Yet none of us, even after several years had passed, said a word about it. Back then we refused to reflect on her passing. Even Peter's odd and blunt comments came to an end, and I didn't like it, not one bit. The silence was deafening. I did my best to conceal my distress, especially in front of George, but I could clearly see the same thing in all their eyes. Our hopelessness was peaking.

A few years after Mary's death, Peter and I pressed George to come upstairs to the activity room, in an attempt to get him to interact with us.

Although he still had his other beautiful sister, Jane, who was not a Thal, the bond he'd had with Mary, Jane was incredible at continue it for all of us but particularly for George. We have all remained grateful to Jane ever since then. Peter had known George years before I'd met them, and it was Peter who managed to finally get George to agree to hang with us. I don't know if he got counselling or any kind of support because he never wanted to talk about that kind of stuff but either way, he was back being himself in no time.

This kicked off a friendship that lasted a lifetime. No matter where we went in the building, we were known by all of the hospital staff as the three musketeers. In the beginning, we were astonished by the attention from everyone who worked at the hospital, shocked that even important people like specialist, doctors and head nurses took time out of their busy days to stop and ask us about our day. The attention we received was so overwhelming some days we were chastised by Sister Shaw for taking too long to returning even from the cafeteria.

Once we became adults, that attention was almost non-existent from everyone working at the children's hospital as we moved casually about the place. The kind name-calling, delightful waves and stares, and even the questions were still there to a certain degree, but for different reasons.

'Mummy, why are these old people having blood. What are they doing here?' We would often hear from little children while we walked together to the cafeteria rolling our IV poles alongside us still attached to our bottle of blood.

The fact that we were still receiving treatment in a children's hospital had a lot to do with our unanswered questions. This remained the elephant in the room that no one talked about, it was all too overwhelming because we didn't want to ask the question, so we all remain silent receiving our regular getting blood.

———

Severino, who was a little older than us, enjoyed going out with us occasionally. However, the three musketeers continued to get together regularly outside the hospital grounds.

By then, we all owned two or three cars. I had a Mitsubishi Sigma Peter Wherrett Special which I'd purchased as new with only a few thousand kilometres on the clock and I loved it. But Peter's car sounded the best. It was a 1971 Ford Falcon V8. It didn't look like much, but it growled like a monster.

George, on the other hand, viewed cars as a means of transport only. He had a four-cylinder Chrysler Galant that was so well-tuned it ran like a breeze. Both Peter and I always took our cars to George for a tune-up in the hope he could maximise each one's performance. He wasn't a mechanic; he was just smart and knew how to do stuff like that. George worked for Telstra as a technician but had a knack of being able to read a manual, buy the equipment and do work as skillfully as any professional.

George was naturally gifted at almost everything. I don't think I've beaten him at chess more than a handful of times. At school, playing chess was my thing and I loved the challenge of playing new people who thought they were good and beating them. Playing George from a young age helped me get to that level, but over the years I could never once beat him convincingly.

Peter, George, and Severino went on many trips and holidays together, but the three musketeers regularly got together for lunch at various pubs, clubs, and bars, each taking turns to choose a place close to where we lived. For Peter, it was in the city area, for George around Bankstown and for me, the western suburbs.

We were all still living with our parents and when we got together, conversations about our treatment were shunned. George was at his shelf-life age of twenty-five and no one was talking about it. Following close behind was Peter and, although he often joked about dying, he didn't dare go into detail. Now and then, we all giggled about it, but mostly we avoided the subject of our mortality, while deep inside we were hurting.

'Fuck me, we are lucky to live this long. What the fuck are you all trying hard to avoid not saying! We have reached our expiry date, all right and big fucken deal! We made it this far so what more do you what, who shout is it?' Serverino belched out on one occasion leaving us all speechless.

Sev felt playing that game was a waste of time. He wanted it over because he was over it. The contempt he showed as he watched us pretend we were not waiting for our deaths was clearly visible, but we continued to ignore the ticking clock and kept our despair silent. Sev pushed the envelope to show us he was not bothered one bit by always talking about it and even mocking us all at times. But maybe he was better at hiding his feelings than we were, and we just didn't notice due to our egocentric youthful arrogance.

One particular blood day, George told us about a girl called Debbie who he had started dating. By then, he'd had many relationships and even nearly got married once but that didn't work out.

Previously, both Peter and I would have loved to get him to talk to us about his relationships, but he was usually a locked vault and not even Peter's best efforts could get him to open up. So when he brought up Debbie, we were all gob-smacked, even Severino, and we patiently listened as quietly as we could. We knew that if we interrupted him, he might never again open up about his private life.

After he had finished talking, we bombarded him with questions about her. In my excitement, I nearly told them about Helen, but at the last minute, I changed my mind.

At that point, Helen was only a desire, a dream. I didn't know her yet and she didn't know me. Nor did I know if I was going to see her again. The week quickly past since we had arranged to meet again the following week, but I didn't know if she was going to show up. Then when she did, I was then hoping after our first date to the movies, did she like me? I was pondering these questions when George called out to let me know my bottle of blood was empty and needed changing.

We all went back to talking about Debbie and that's when George decided he was going to bring her in on our next blood day. *There goes our regular lunch get-together,* I thought. Then an icy cold shiver rolled over me. *Even if Helen ends up meeting me next week, it will be over after I tell her all about my need for regular blood transfusions.* Not to mention that I was going to die soon.

I can't believe it, I thought. *Finally, I meet*

someone special and it's over before it starts. My life is in ruins. The reality of my chronic blood disorder made me hit rock bottom and my heartache and despair grew and transformed into anger. I was angry for all of us, angry that none of us would have a chance to live a normal life. *It's not fair!* I was about to scream it out loud when I got a hold of my emotions and managed to fight back my tears.

21

THE ENGAGEMENT, 1984

In the days leading up to our rendezvous, my concern that Helen wouldn't turn up was intertwined with a strange feeling of relief at the thought that perhaps it was best if she didn't come. Although I was going out of my mind, desperate to see her again, I didn't know what to do if she did turn up. Telling her about my condition was a must, my problem was when to do so. I found myself praying that she wouldn't turn up so I could be spared the grief of losing her.

I believed that if I told her early in the relationship it would be easier for both of us. If I waited too long to tell her, I would likely lose her trust and never win it again. So I decided to get it over with.

My torture was over upon seeing her pull up in

her HQ V8 Holden sedan. She looked spectacular in it and my mouth remained open while she parked her beast of a car. But when she got out and jumped into the passenger side of my car, my heart raced with a sense of joy, and I forgot everything as if I didn't have a worry in the world.

After greeting one another with a smile and a polite chat, I drove off with a grin plastered to my face and left all my worries behind. She had turned up! I was beside myself with joy and happiness thinking that maybe she liked me. This thought was too consuming, and I forgot what I was worried about as we drove for hours, talking all the while, before stopping to get something to eat.

Having worked for a big bank after graduating from high school, Helen was now working for the Health Insurance Commission, which explained the car she was driving.

'So how is it like working in the family business, takeaway is that right?' Helen asked sounding sincere and interested.

'You remembered, yes it's good but I'm not good working with my bothers. The business is great but really long hours.' I said thrilled Helen was interested.

It wasn't until a few days after our rendezvous that I remembered what I needed to tell her but, by then, I didn't care because I couldn't stop dreaming about her. My every minute was consumed thinking about her. Dreams about her filled my days and nights.

Weeks of meeting regularly turned into months, and I still hadn't said a word about my condition. I loved her now and I hated myself for not telling her for fear of losing her.

Then at the start of our third date, I stopped the car on the way to dinner and finally belched out everything about my condition. I didn't leave a thing out. I made sure I was thorough. Considering that I might not see her again once I told her, I took my time.

'Did I talk too fast; do you want me to repeat anything?' I kept on repeating over and over again.

Once I'd finished, she smiled at me and didn't ask a single question. We just carried on talking about our usual things like our work, family, and general conversations about our favourite TV shows. I was beside myself with shock at her response or lack of it. I couldn't keep up with our conversation after that, but I didn't dare interrupt her. After a while, I slowly got back into the conversation, and we spent the rest of the night talking before I took her back to her car.

She didn't mention it on any of our following dates, either. This astonished me. I soon realised she was a rare and beautiful gift, the kind you came across only once in a lifetime. A gift so unique that, if you weren't looking, it could easily pass you by. But I'd found her, and I saw she was gorgeous both inside and out.

Soon afterwards, I told my family about Helen, and she also told her parents about me but left out my condition for now. My parents, although happy for me, warned me to not get too serious with her and to not tell her about my condition. They were conflicted. On the one hand, they loved that Helen was from a Greek family, but on the other hand, they felt I couldn't let the relationship blossom because I was sick.

'What happens when her family finds out you have blood, no good no good.' Mum shouted out so loud that the entire street could hear her.

'Shout up mum, it my business. Dad, tell her it's my business.' I screamed back at the two of them.

'Don't say shout up to your mum okay.' Dad said before turning to mum and telling her to shout up himself.

But I had made up my mind. She was the one for me. I couldn't live without her. Ignoring my parents, I insisted they cease their ridiculous opposition by telling them that if they didn't listen to me, I would move out of home. Worried I would follow through on my threat, they both backed off.

I finally persuaded Helen to organise a time for

me to meet her parents. Her family didn't like not being in control and dreaded the thought of her bringing home someone who was not Greek. Her sister Voula, two years her senior, had been married one week before her eighteenth birthday to a Greek guy, Steven Hatzimanolis, and the fact that Helen was twenty and still single was a bone of contention between her and her parents. She had persistently rejected all her parents' attempts to arrange a match for her with a Greek man hand-picked by them.

The day finally arrived. While walking up the driveway to Helen's family home, she and I were confronted by her two younger brothers. Contrary to her warnings, the boys were both approachable, friendly, and remained polite. After a brief meet and greet, Nick, who was less than ten months younger than Helen, said he recognised me from Mount Druitt High School. He was in the same year at school as Helen. Con, who was two years younger than Nick, couldn't recall seeing me at all.

'Hey man, good to meet you.' Nick said with a genuine smile.

'I like your car, great colour man.' Con said while pointing back at my car.

Thanks guys, nice meeting you both. I think we better go in now.' I reply shaking both their hands.

I couldn't recall ever seeing Helen or her brothers at Mount Druitt High either. Although I'd had some awful and shocking experiences at school, after hearing some of Helen's stories, I think she had it worse by far. School had been a terrible time for both of us. Both Helen and her brother Nick had been in the year below me and I couldn't believe I'd never noticed her. It would have been grand to have one another during those difficult times, sharing and caring during the turmoil.

Helen opened the screen door and we walked in. After taking one look at me, Helen's mum abruptly said to her dad, 'Jonny, Πήγαμε στον χορό.'

Upon noticing my blue eyes and thinning blondish brown hair, Helen's mum assumed I was an Aussie (I am, actually). However, I clearly understood not only her Greek words which translated to, 'Johnny, we entered the dance', but also her meaning—that she was in no mood for a party with a non-Greek.

Their unhappy countenances rapidly revealed shock then joy, followed by grins from ear to ear once they heard my greetings to them in Greek. Their transformations were miraculous. They took me by the hand and sat me down and bombarded me with an array of food and drinks.

Once all the pleasantries were over, Helen's dad didn't waste any time. He turned to me and said, 'What are your intentions regarding my daughter?'

Although not unexpected. And I had been considering how I would reply to such a question all week, I was unable to give him an answer. Because, once Helen heard her dad ask me that question so early in our meet and greet, she grabbed me by the hand and dragged me outside to my car, nearly dislocating my arm and shoulder in the process.

Helen had her reasons, and I didn't dispute them but I would have enjoyed replying to him on that occasion. He asked me again much later and after he'd met my parents, but it wasn't the same because we had grown on each other by then, and he knew what my answer would be. It was the beginning of a very close relationship and I found myself becoming comfortable with both Helen's parents and able to talk to them about everything. Well, almost everything.

In May 1984, we officially announced our engagement. This was a huge relief for Helen's parents but earth-shattering for mine. My parents disguised their distress well and didn't know how to tell me about their concerns. But it boiled over one night after I overheard them talking. I told them that it wasn't their burden any longer and asked them to respect my decision.

'But her parents will make her leave you once

they find out you have blood.' Mum would say be-fore having dad repeat the same thing verbatim.

'So what, she can leave any time she wants. If Helen decides she want to leave, I will support her because it's her choice!' I shouted back with a smile.

Once I explained to them how Helen and I felt about each other and that thinking the worst wasn't helping me at all, they both started to do something they had never really done before, and that was to listen to me. It was strange to them but, after hearing me out, there was nothing they could do. Their fears of Helen leaving me after finding out I was sick didn't faze me at all. They didn't know that, just as George had brought Debbie into the hospital on one of our blood days, I had already taken Helen in on several occasions. I had told her everything I could about my condition, and she didn't falter, not one step. I had opened up to her and not left anything out, and she had been incredi-ble. She made me feel as if I was normal, and that I was fearless with her by my side.

A few weeks into our engagement, Helen called me late one night just after I had put on my Des-feral pump. She excitedly informed me her dad had won a small amount in lotto. We talked for a while until her dad told her to get off the phone so he could call Greece.

'Wow, so happy for your family. If anyone deserved it, they do sweetheart.' I quickly said after hearing her dad wanting to use the phone.

'Thanks honey, I wanted to tell you first. Talk to you tomorrow, bye honey.' Helen replied before quickly hanging up the phone.

Once all the excitement was over, we found out he had won a very small portion of the big prize that was enough to pay off their house, cover a family trip to Greece and maybe provide enough to buy a house in a nicer area, if they didn't spend too much on their Greek holiday.

Helen was so happy for her parents. 'They've been hard workers all their lives,' she said, 'they deserve this win.' I loved seeing the joy she had for them. Her parents offered to pay for a party to celebrate both their win and our engagement before the family headed off on their Greek holiday. We were delighted to accept and hastily prepared for the event. My parents feared the worst, but we grabbed hold of it with both hands and ran with it.

The party was held was in Earlwood in a venue that was a mix of Italian and Greek styles. We had a live band and a master of ceremonies who called me Harry all night. Members of both families were introduced one by one, just like at a wedding, and from the very start, the MC introduced us to all the guests as Helen and Harry. The party was a whirlwind of hysterical laughter that continued the entire night, with speeches, magnificent food, and dancing.

The entire time, I couldn't stop staring at Helen. I still couldn't understand. Why had she picked me? I lived with and worked for my parents. I had no trade or professional qualifications and, what's more, I was going to die any day. It baffled me. She, on the other hand, was an absolutely gorgeous young woman in a professional job, who owned a beast of a car and was bursting with goals and aspirations. What was I doing with her?

But this was our night, and I was determined to enjoy it.

22

THE WEDDING, 1985

Helen sent word that she was returning home alone from her trip to Greece and asked me to pick her up from the airport. It was two months into their three-month family holiday, but she needed to get back to work and the excuse worked.

The shock on my face upon seeing her for the first time after two months took her by surprise. From my reaction, she thought I didn't want her anymore. Instead of grabbing her, lifting her in the air and turning her around, and telling her how much I'd missed her, I found myself pulling back. I couldn't believe she was there in front of me, and I didn't handle our reunion well. Seeing her tanned, trim, and looking even more terrific than ever before, I was gobsmacked.

What the hell am I doing with this beauty? I'm going to stuff up her life and she doesn't deserve that. I felt myself sinking in a deep, dark hole of despair. She was so incredibly beautiful. How could she possibly be mine?

Just when my heart was breaking, that's when Helen did her thing, and I was Iron Boy once again. 'I love you,' she said. 'I missed you! Aren't you going to give me a hug?'

Her words snapped me out of my anguish, and she fell into my arms. Once she was there, I was superhuman again.

The year was nearly over, and we set the date for our wedding for early November the following year. Both sets of parents were excited for us and when they asked us where we were going to live, we just smiled. We had no idea; we were just focusing on the wedding.

In our free time, we drove around to look at all the local house and land packages for sale that seemed to be springing up everywhere. I told Helen all about my brief career in selling real estate after leaving school and gave her my opinions on what was on offer.

It was on one of these weekend drives in a new housing estate called Erskine Park, that we came across an offer we couldn't refuse. To purchase a block of land in the estate, all we needed was a fifty-dollar holding deposit and then we had six months to pay the full deposit. This changed everything for us. We had no money saved and buying a house or block of land was the furthest thing from our minds. Buying a home was merely a dream for us. Viewing houses and land for sale was something to fill in our weekends, something we enjoyed doing together. But this offered a real possibility for us to have something of our own and a way to avoid renting.

We knew once we started renting there would be little chance of us ever saving enough money to buy a house of our own. All we needed to do was raise the deposit within six months, borrow just enough from the bank to purchase the land, then select a builder that suited our budget to build our house, hopefully before we got married.

To our shock and dismay, after placing the holding deposit down, both our families laughed once we told them we'd purchased a block of land for fifty dollars. We explained that the holding deposit gave us six months to save up the full deposit, but everyone said we didn't know what we were talking about and ridiculed us as ignorant individuals.

'They tricked you stupid, you must be both dummies. Jesus Christ, they took your fifty dollars. Don't you understand nothing, oh my god?' Mum screamed out while waving her arms up and down.

'Calm down, why are you screaming like that and why are you so upset?' I replied with disbelief seeing them both that way.

Six months later, after successfully getting a home loan, we purchased the land and finally selected a builder. Then work commenced on our three-bedroom, one-garage castle. It was good to see the shocked faces of our family members who were dumbstruck at our accomplishment.

At first, we couldn't make it work, but we pushed hard for it, we were able to secure the money to purchase the house and land by getting two loans from two separate banks.

It was costly, but that was the only way we could make it work, given that we had no savings. Back then, fifty thousand dollars was an enormous amount of money, but we worked the problem and came to a solution. It was satisfying to know we'd done the right thing while our home was getting built, as we came up with the huge monthly repayments.

With our castle on schedule to be completed days after the wedding, our focus turned to the wedding itself. This event was a complete contrast to our extravagant engagement in every way, as money was an issue. We didn't care because we were looking forward to moving into our new house once we returned from our honeymoon. To save money, we hired the local YMCA hall and a DJ rather than a band, and Helen hand made each of the magnificent bridesmaids' dresses.

Despite the low-budget wedding reception that was ahead of us, I eagerly awaited Helen's arrival at the church. She took my breath away that day. She was simply stunning. *I must be the luckiest guy in the world,* I thought, *to have someone that smart and beautiful love me so much she wants to marry me.*

My best man, Andrew Patsos, was a beautiful man whose love and support encouraged me to become the man I needed to be to start a family. He was brilliant just having him around. He was my mentor over this period and a wonderful friend who freely gave me much needed hints and nudges, pointing me in the right direction when I needed guidance leading up to our wedding.

Ten years my senior, Andrew has been a family friend and knew about my condition over the years. Married to Christina, they had three beautiful children, Vicki, Joanne, and little Con. Although a family friend, we had gotten to know each other much better on a plane trip to Greece when we found ourselves on the same flight. Andrew had been travelling alone to visit his aging mum who was ill, and I was with my mum and two younger brothers, Con and Angelo.

It was on this trip, after spending some time

with Andrew and his family in Greece, that we made a pact with each other. This was after a discussion we'd had sitting in a hillside café overlooking the magnificent Gulf of Corinth sea in the village of Akrata.

'You have three beautiful children and a wife who all love you very much, I don't even have my health. Why do you say I'm lucky?' I asked sitting opposite him at a half-filled tavern in his village just before ordering my meal one late hot evening.

'You listen to me, okay! Nothing wrong with you, okay! You lucky because you are young, okay! Also you too will get married and have kids, okay! Trust me, okay!' Andrew replied firmly and with much conviction before ordering another drink in his beautiful heavy Greek twang.

'Okay, if I ever get married, you can be my best man.' I quickly answered back hoping for a quick response.

'I've never been Koumbaro before, me Koumbaro. Okay, put it here!' Andrew yelled out before jumping to his feet to shake my hand then almost pulling my arm out of my shoulder while also lifting me completely up and out of my chair.

That day, Andrew predicted I would get married. He gave me his assurance then and there, that he would certainly be my best man. I was so happy, but secretly I thought he didn't know what he was talking about and dismissed his words as just talk although my arm was still hurting from his poised prediction.

Five years later, it was now late 1985, and we were having our very own Greek wedding. Andrew stood next to me in the Greek church at St Marys in front of family and friends, watching Helen walking down the aisle with her father.

Andrew was incredible that day. All my other friends and family had gotten to know him at our fancy engagement so by then, everyone knew him well. Boasting he had never been a best man (Koumbaro) before, everyone could see he was just as nervous as I was, which made him even more endearing.

In my experience, Greek weddings were long and arduous, full of traditions and ceremonies of epic proportions that tested even the most tolerant of people. But when it's your wedding things are totally different. As soon as I saw Helen's eyes, I was lost, mesmerised by her beauty, and I wouldn't have cared if I was standing there naked. You could have got me to do anything at that moment. She was so beautiful.

The next thing I knew, we had cut the cake and were taking off in the limousine and heading for the airport to start our honeymoon at Surfer's Paradise in Queensland. I had floated through the day in a blissful dream, and I even thought our wedding reception hall wasn't too bad I guess. Well, at least I didn't notice!

It wasn't until we got back, and I saw the wedding reception photos that hit me like a tsunami. Once I had taken off my rose-coloured glasses, the thriftiness of our wedding reception was obvious, and I wondered if maybe that had been a mistake. But getting into our own home made everything all worth it.

23

OUR NEW HOME, 1985

As assured by our builder, Rosewood Homes, the day we arrived back from our honeymoon in late November 1985, our new home was completed and ready for us to move in. It looked even better than we'd imagined. Nestled on a flat, above-average size block of land, the pretty, three-bedroom, single garage structure was beautifully finished.

Settling into Erskine Park was amazing, but it was strange being only one of half a dozen homes built and occupied in our street. One downside about the neighbourhood was the constant yelling that came from the house across the road. From the day they moved in, their arguments occurred like clockwork every evening, and we would hear them screaming at each other for a couple of hours, growing fainter into the dead of night. We assumed they were newlyweds like us, but I dared not introduce myself, fearing a negative response.

Before we knew it, we were expecting our first child. On the evening before the birth, things became rather difficult. On a positive note, this helped to take away the fear, but it also robbed me of the joyful anticipation.

I had arrived home late from work after a hectic and difficult night, early in May 1987. I remember it clearly because I had almost been fired after my boss had busted me unintentionally neglecting my work responsibilities.

My parents sold the family takeaway business not long after meeting Helen. It was a while also before my dad finally got over the shock that I had got off the disability support pension which was termed as the invalid pension back then without telling him. Even after several good jobs, my dad still had a hard time dealing with my decision.

By now, I had been employed by Kentucky Fried Chicken only for the past year, working as assistant manager in a KFC outlet at nearby suburb of St Marys, and I was torn between my responsibilities as a duty manager and those of a concerned husband and father to be. Before this particular day,

I had been working extremely hard towards becoming the store manager. I knew I was ready for the promotion.

With our first baby due any minute, like any prospective father, let alone one who had not expected to even grow old enough to have a baby, I was excited and distracted. During the afternoon shift, I had spent most the time on the phone to Helen instead of doing my job. Exhausted from reassuring her and on tenterhooks as the baby could come any time, I was looking forward to the end of the shift.

I found out later during the night that word had got around that a new area manager by the name of Stephen was conducting store checks, as well as simultaneously meeting all the afternoon duty managers. He intended to evaluate and seek out for himself the best assistant managers. Apparently, he was looking for good-quality people to promote to store manager positions.

Unfortunately, as I was on the phone to Helen most of that night, I didn't hear anything about these checks. I was unaware of the situation and, to be honest, I didn't care much by then, I just wanted to go home. When the new area manager entered my restaurant, it was obvious his initial impression was far from favourable. In fact, he looked appalled and aimed an almighty death stare in my direction. After giving my store the once-over, pen in hand, he took his time to finish writing his notes before giving me his thoughts.

Luckily, the store was empty of customers, and I was astonished to find it was nearing eight pm. Where had the time gone? The new area manager had observed my staff working hard but I now noticed the sheer terror written on the faces of each staff member. It occurred to me that perhaps it would be a good time to hang up the phone before the area manager, in his anger, seized it from my hands.

His face livid with rage, he strode towards me. I braced myself for the worst. He confronted me, intimidating me by invading my personal space. With disgust dripping from his voice, he abruptly requested my presence for a private chat in the front of the store. I gathered what little pride I had left and complied. As I led him out, I took a deep breath for courage but felt faint-hearted and ashamed for letting my staff down.

Outfitted in the company uniform and a blue name tag, the tall, well-groomed, and serious-looking area manager was so inflated with anger I was afraid he would explode all over me. I could almost imagine steam coming out of his ears like a volcano before an eruption.

He was so angry, at first, he didn't know where to begin. But when he did, his furious words poured out of him like an endless lava flow. I sat there in silence as he ripped me apart. I felt diminished, torn into pieces like a sheet of paper emerging from a paper shredder.

Repeatedly, during this humiliating lecture, he declared that any attempt by me to respond would be futile. He insisted nothing I could say would get me out of this mess, as he had just witnessed for himself my overall lack of management skills, resulting in my store being rated the worst by far of all the stores he had inspected that night. 'You'll never make store manager in this company,' he said. 'And forget about ever getting another pay rise.'

Concerned about Helen even though I had no reason to do so after she reassured me, but my mind was still on her to fully take in his diatribe.

At one stage, I tried to clarify a few points and salvage some dignity, but my attempts were a miserable failure. So I just sat there watching his lips move but without hearing what he said. I couldn't wait to go home to Helen and get some sleep. I was tempted to get up and walk out but decided to wait it out. I was sure I could somehow explain the situation when senior management knew the facts.

Eventually, the harangue was over. He handed over my first written warning, arrogantly reminding me that I only needed two more warnings before being terminated. After superciliously advising me my days in the company were over and I'd be better off resigning, he turned on his heel and stormed out of the store.

With all hopes of ever making a good impression or getting a promotion shattered, I lingered for a while, just sitting there all alone. Then I took another deep breath and picked myself up to initiate the closing procedures.

The staff that night were extraordinary. They worked so hard without saying a word to one another or me. They just cleaned up, working faultlessly together as a team. Their silent sympathy was soothing but I couldn't help thinking that I had not only let myself down but, more importantly, I had let my staff down as well. I felt as low as I ever had.

We finished up in record time. As I escorted out the last member of staff, I was thankful we'd had no customers there to witness my humiliation. The last staff member hung around silently until I locked up and switched on the alarm before he got into his car and drove away. I managed to get to my car without topping off the night by becoming another gang-related street attack statistic. It was a calm and peaceful night out on the streets, in contrast to my internal distress.

By the time I got home, Helen was sound

asleep. Determined to get a good night's sleep after my traumatic day, any thoughts of using my Desferal pump were quickly discarded. I noticed it was just before midnight as I quietly slipped under the covers and fell into a deep sleep.

24

JIMMY, 1987

When Helen got up about 2.30 am that morning with labour pains, I slept on. Due to a couple of false alarms and trips to the hospital a few days earlier, she thought she would wait out the pains this time by doing the rest of the ironing and other housework. But as the hours passed without any relief, she knew this time it was different.

With her labour pains increasing in severity and frequency, Helen got changed, avoiding waking me until the last possible moment. The clock on the bedside table showed 5.30 am as her initially gentle nudges became progressively more forceful. 'Wake up, wake up honey, its time!'

In an instant, my bellowing snore became a crackly, irritable voice. 'What? It's time? Oh, it's time... the baby!' I screeched, almost deafening my wife, and probably waking up the neighbours as I kept on repeating it over and over.

Panic-stricken, I tossed the bed covers up in the air. As they came back down they completely covered the mum to be who was still sitting on the bed, scared out of her wits and flabbergasted by my response.

'Helen, it's time, it's time, it's time,' I yelled.

In my hurry, I tripped over myself attempting to get dressed, only to be disturbed by a howl from underneath the bedcovers. 'Sorry honey, where are you?'

As I burrowed carefully down beneath the pile I had created, suddenly the bedcovers were tossed back.

'Sorry, honey. Are you sure this time?' I asked.

'Yes! Yes! Yes! Let's go now, please!' Helen screamed.

Tenderly assisting my wife up from the bed, I cautiously held her with all the strength I could muster while being aware of her intense pain as we both took one step at a time.

Upon reaching the end of our bed, I helped her sit down for a moment to catch her breath while I finished getting dressed.

'Arthur... Arthur... I've got to tell you something.'

'Yes, Helen?'

'When I finished packing my overnight bag, I looked out our bedroom window and I couldn't believe what I saw.'

'What honey?' I replied, still madly dressing.

'Butch (our dog) is on the roof of the neighbour's car across the road fast asleep!'

'What? Not the neighbour directly across the road?' I asked, my voice shaking.

Finally dressed, I looked out the window to see for myself. Butch, our huge two-year-old German shepherd, was out like a light on the neighbour's car roof with all four legs dangling over the side.

With my thoughts full of the possible reasons justifying why our neighbour's wife had left her husband after another neighbour's recent confessions to Helen confirming he was a wife beater, I whispered, 'Get down, Butch'.

'What are you doing?' asked Helen. 'You'll have to go over and get Butch off the car before that man goes to work.'

'Bullshit. You know he's a psycho. If he catches me getting Butch off his car I'm dead. Let's go before he comes out. Butch, get off that car!'

'Look, you know he goes to work early, so you'd better go over there before he comes out. I'm not leaving until you do. I'll be fine. So go and hurry up.'

We both nervously made our way to the front door. I had a tight grip on Helen and had become aware of her furiously trembling body. I was convinced that this really was it. Our baby was coming!

After making sure Helen was okay and scared to death that my neighbour would come out any minute, I quietly yelled, 'Butch, come here, boy.' There was no response. Keeping a vigilant eye out, I reluctantly walked over to the neighbour's car. Butch slumbered on in guiltless bliss. As far as I was concerned this wretched dog seemed determined to get me killed. Even when I was right beside him, Butch was still oblivious to my presence.

In a last desperate attempt, I shouted almost into his ear, 'Butch!'

The stupid dog finally lifted his head, turned, and looked me in the eye with the biggest bloodshot eyes I've ever seen. He had no concern at all for my predicament and, with his trademark stupid doggy grin plastered on his face, he looked up as if to say, 'What do you want?'

'Butch, get down now!' I yelled.

He finally realised I was serious and tried to raise a smile as his tongue flopped out of his mouth and his tail slapped the windscreen.

By now, dawn was breaking, which increased my panic. How was I going to get Butch off the car? I repeatedly whacked him on his backside. Ignoring me, he casually stretched each muscle and extended his frame over the top of the car. I couldn't believe it as I stood there in clear view of the whole street with the night breaking into a beautiful, brilliant blue and red sky.

My embarrassment continued as Butch proceeded to slide down the car windscreen backwards, leaving muddy tracks from all four paws and other smears of unmentionable origin. He finished up by leaving great scratches all over the bonnet as he lost his grip trying to jump off. With a wave of his tail, he slowly toddled towards our house continually looking back smiling at me with his tongue hanging out of his mouth.

I stood there undecided between leaving the crime scene or confronting my worst fears and telling my new neighbour about Butch's destructive night's rest. Still exhausted from the previous day and having had little sleep, everything seemed to be moving in slow motion like a nightmare. For a moment, I hoped I would wake up and laugh it all off. But it was all too real.

I heard the neighbour's front door open. Leaving the scene of the crime was now out of the question. I turned to face my punishment like a man, standing next to my neighbour's almost new white station wagon. It was now considerably less new since Butch had made his unique additions. I'd never really spoken to this man; I didn't know even his name. I just knew he and his wife, who had previously screamed at each other every night, had recently separated.

'Excuse me, can I speak to you please?' I asked hesitantly.

'What?' he asked. He didn't seem surprised to see me in his driveway.

'Well, I'm deeply sorry about this, but my dog slept on your car roof last night and scratched it. I'm so sorry. I'll pay for the damage, so don't worry.' All the while, I was very conscious that poor Helen was waiting for me to take her to the hospital.

The man didn't say a word, just walked right up to the car, gave it a quick once over and turned away. 'Look, mate, don't worry about it.' He got into his car and drove away.

Relieved, I ran back to tell Helen. She had managed to walk outside, her contractions still slowly increasing in intensity and frequency. I helped her into the car, stowed her bags, and locked up Butch and our house. On the way to the hospital, between Helen's labour pains, I told her about the lucky break I got from our neighbour.

'Did you get his name?' Helen asked between breaths.

'No, and I hope there's not another opportunity to do so,' I replied.

On arrival at Blacktown Hospital, I was confident Helen would be able to walk to the maternity ward from the car park. So I drove past the emergency entrance into the car park and carefully parked the car. We looked at each other, both speechless for a moment or two. Horrified, we were no longer convinced that Helen would make it to the maternity ward under her own steam. I helped her out and locked the car, leaving her bags behind. Helen placed her arm around my neck, and we proceeded to walk along a brick wall towards the maternity ward.

But we didn't get very far before she was overwhelmed with a massive contraction. It was so forceful I buckled like a wet paper bag almost falling to my knees as I tried to hold her up. Her contraction continued stronger than ever as I tried to hold her up against the wall.

'Arthur! Arthur!' she kept on calling, even though there was little I could do.

I panicked and told Helen to press her back up against the wall while I went to get help. She didn't reply and my panic turned to confusion. What was happening? Suddenly, out of nowhere, a nurse came towards us with a wheelchair. We were both exhausted and had no idea how long we had been struggling.

The nurse gently assisted Helen into the wheelchair and wheeled her off to the maternity ward. I was too useless to do anything other than follow them. Before they vanished behind opaque plastic double doors, the nurse instructed me to go and get Helen's bags.

Relieved my wife was in good hands, I took my time getting her bags while trying to catch my breath. Returning with bags in hand, I noticed the nurse coming out of the no go zone looking serious. She approached me and asked me to take a seat while they prepared Helen for the birth. She said she would call me soon so I could get suited up for it as well.

Still holding Helen's bags, I sat down on an uncomfortable chair. I stared blankly at the doors, unable to believe the last twenty-four hours. I realised with awe that we had reached the moment of truth. Then I started to have dreadful thoughts about how the baby would turn out. Would he or she be healthy? Would the baby have Thal? It was unbearable.

My thoughts drifted back to Helen's pregnancy. I recalled our initial excitement and then plummeting to the depths of despair during our horrifying first consultation with the obstetrician when Helen had walked out in shock after being advised to terminate her pregnancy. Thankfully, after we had sought a second opinion, the new paediatrician had put us back into a positive and calm state of mind. We knew the odds and what outcomes to expect, but I couldn't help thinking that no one ever really knows for sure. Helen didn't carry the Thal gene, but that knowledge didn't give me much comfort.

I felt nauseated and was trying to catch my breath when the nurse appeared. 'Mr Bozikas, your wife is ready for you.'

For a moment I started looking around for my dad, and then I realised she was talking to me.

'Here, put on this gown and put your wife's bags over here next to the bed,' she instructed with a gentle smile.

Before I knew it, the action had started, our paediatrician had zoomed in, I had cut the cord and behold, we were parents! Jimmy was born in early May 1987, weighing seven pounds (3.1 kilograms) and 52 centimetres long. And he didn't have Beta Thalassaemia Major. In no time, I found myself holding Helen as she held our new baby boy. I couldn't believe my eyes. Then I prayed a small prayer thanking God for our little blessing and for still being alive to see this miracle.

The paediatrician slapped me on the back. 'Everything is fine, DAD!' he said and walked out.

Reassured, I stopped counting Jimmy's fingers and toes. Everyone left and gave us a couple of minutes alone. It was breathtaking and exhilarating to watch this little bundle of joy snuggled up smelling fresh and bursting for affection.

The nurses soon came back and took Jimmy to clean him and then placed him in a small cot next to Helen. His cry sounded like angels singing, which left us speechless and looking at each other with pride. Once he was snuggled up in his cot, the nurses freshened up Helen.

After a few hours of absorbing our new baby's glow, it was time to notify the rest of the family of the arrival of our healthy baby boy. Hoping to give Helen time for some well-earned sleep, I staggered out of the maternity ward and was blinded by the bright mid-morning sun. Still on a high, I somehow found a phone and started calling our families.

'It's a boy! It's a boy!' I repeated over and over again. In my excitement, I forgot to tell them we were at Blacktown Hospital. Instead, I recited the precise birth time and Jimmy's weight and length before hanging up. *I am on fire*, I thought.

Even though things hadn't looked too good for me at work the day before, I phoned my manager and told her where I was and requested a week off, starting immediately. She already knew about the situation with the new area manager but was delighted to hear about the birth of our baby boy. She put two and two together and realised the pressure I had been under the previous night and, to my relief, agreed to my request for time off.

On cloud nine and bursting with confidence, I strode back into the maternity ward. But the delivery room was empty. Helen and Jimmy were gone. I raced outside and found the nearest nurse. 'Where are my wife and baby?'

'Sorry, Mr Bozikas, we moved them into the ward with the other new mums,' she replied cheerfully.

I ran to the ward, gasping for breath, and found them both fast asleep beside one another. I didn't have the heart to wake them. I just stared at them in wonder.

For the first time, I came to grips with the reality of Thal and what it meant for me and my new family. Although I'd managed the time-consuming and sometimes painful treatment, I'd never thought about my mortality since the day I was twelve and had decided to live. Now, Helen and I had brought another new life into the world. But how long would I be in Jimmy's life? How long did I have to live? The odds were against me. How would Helen manage if something happened to me? These questions were not new, but now they were ever so more important.

———

My best man Andrew's astuteness and intuition were revolutionary. When it was time for Jimmy's baptism, he decided to have someone assist him with christening our first child. When he got his eldest daughter Vicki, who was not much older than ten at the time, to assist, it was priceless.

On the day of the christening at the Greek church in St Marys, surprisingly, there were no arguments from the priest, and it went off like clockwork. Vicki took the lead once her father executed his responsibility of christening Jimmy and took everyone's breath away with her confidence and astuteness.

Father and daughter teaming up was viewed by all to be so synchronised and polished, it really did look like they were paid professional doing it on a regular basis. It wasn't until after all the huff puff was over, they both revealed how terrified they both were going over each step of the way and it was the energy they got off each other that got them through it all.

Even though there was confusion around at the time from everyone about why he'd asked his daughter to assist him, he wanted to share the beautiful moment with Vicki. He valued his family and loved imparting responsibility to them.

Wonderful things were happening in my life.

25

PAMELA, 1988

Surprisingly, twelve months later in early May 1988, Pamela, our second bundle of joy arrived. She weighed seven pounds nine (3.3 kilograms), was 51 centimetres long and, to my intense relief, also did not have Beta Thalassaemia Major.

Although more prepared this time around, I wasn't ready for a baby girl. I had grown up with three brothers, so I knew nothing about little girls. What was I going to do with her?

As with Jimmy, when we first heard our new bundle of joy in the delivery room, our hearts opened up and filled with overwhelming joy. Our daughter was gorgeous, and she melted our hearts from the beginning. We were left speechless, gazing at her for hours.

By then, Jimmy was crawling and once they put Helen and Pamela in the maternity ward, we couldn't get him out from under Pamela's cot, which was next to Helen's bed.

With two babies at home, life was incredibly hectic. Although I didn't know it at the time, this busyness was a huge blessing because it took my mind off my mortality issues. Helen's hands were so full, I didn't know if I was Arthur or Martha at times just watching her effortlessly go about her daily routine. No longer did she have any free time for herself, and I was in awe with her natural abilities of motherhood.

When those dark defeatist thoughts descended on me late into the night before I fell asleep, suffocating me with fear, Helen reassured me with a wise word or two. That's all it took, and I would be back on track. Right from the beginning, whenever I derailed, she would put me back on track by saying, 'How do you know when you are going to die? Only God knows! I could get hit by a bus tomorrow and go first. One never knows!'

Her insight left me dumbstruck every time. She had a talent for doing this for me right from the start and has remained the light in my darkness.

Once Pamela started crawling, I convinced Helen to add another seat to my road racing bike so I could take both kids with me around the block from our house. The bike was new and, although it wasn't made to take any load, front and back seats to carry small children were available for sale. Jimmy looked forward to a ride whenever I had a day off work and soon Pamela nagged me to take her along on our regular rides.

Once the second seat was installed, I secured Jimmy in the front seat on the handlebars and strapped Pamela into the rear seat. We all gave Helen a big wave and took off on our ride. At the beginning of what would become a regular bike riding adventure, I won both the kids and Helen's confidence with a series of short five to ten-minute rides.

Things were going well for a few months with both Helen and the kids now looking forward to our rides as much as I did. Then one Sunday afternoon, as I pedalled furiously down a local gravel road, I realised my confidence in my skilful balancing techniques at high speed had crossed the line to complacency and carelessness.

I slowly came to a stop and dismounted. With my legs shaking and my heart pounding in my chest, I pushed the bike with the two kids still secured in their seats all the way home. They didn't mind—they sat back without a care in the world, enjoying the sun and our surroundings. Incapable of holding a conversion, I occasionally pulled funny faces and they giggled at me.

For quite some time, my legs trembled, and my heart raced as if I was still powering along on the bike, racing like the wind. It was not from exertion, but from fear, as I grasped the possible implications of riding at ridiculously high speeds on a gravel road with both my kids, who had no protection whatsoever. Just the thought of what could have happened made it hard to breathe.

How could I have been so irresponsible? I wasn't riding with them in a park at a walking speed, I was tearing up and down our local streets, dodging cars and other vehicles at breakneck speeds and not even slowing down when the road surface turned to gravel. I wasn't a dad; I was a moron!

When I got home, after I carefully taking both the kids off the bike, I removed the back wheel and placed the bike and the wheel in the shed. I decided that when Helen and the kids next hassled me about taking them for a ride, I would tell them that we couldn't ride because I needed to fix the flat tyre. It would satisfy them for a while. I didn't want to look at the bike ever again and tried to stop thinking about what could have happened if I'd had an accident.

I knew a lot about accidents, having owned two

motorbikes before I'd married Helen. My entire family, including all my Uncles and aunties, had pleaded with me not to get a motorbike, so I'd bought two. The first was a Suzuki 175cc road and trail bike and the second a Suzuki 450cc road bike. They were both registered to me, and I loved riding them. Off-road, I explored tracks and trails everywhere possible with the trail bike, then swapped to my road bike to visit friends and family or just go for a ride.

But no matter how good a rider I thought I was, people couldn't see me on the road, and I was forever having all sorts of accidents. Barely a week would go by without some kind of minor mishap and, on too many occasions, a more serious accident, resulting in my parents picking me up from hospital after a day or two's stay. I hadn't cared because I felt my life was going to be over soon anyway. *What the hell*, I'd thought, *I'm going to enjoy life as much as I can!*

What was probably my worst accident happened after I'd got engaged to Helen. Straightening up from a left hand turn I collided in the driver's door of the sedan that was doing a U-turn without indicating. My motorbike remained in his door but after my body flew over the car, my head first followed closely then by my body, slammed on the ground hard on the other side of the sedan.

Afterwards, I woke up in a hospital ward and

the guy in the bed next to me had a cage over him with a white bed sheet covering him from head to toe. When I finally got the all-clear to go home, I nearly passed out after seeing the condition of my neighbour. The white sheet covering him had slipped to one side, partially exposing his injuries. Gravel rash from a high-speed motorbike accident had left him with almost no skin on his entire body.

The realisation that could happen to any motor-bike rider left me shell-shocked. I couldn't afford expensive protective riding gear. I finally agreed with Helen and my family to sell both my motor-bikes. From the first time Helen saw my motorbikes, she made it clear to me on every occasion she loathed motorbikes with a passion. I avoided talking about them with her because it only caused her much distress. It was the best decision I ever made. I had enough to worry about without adding more things to my list.

———

When the kids reached their toddler age, Jimmy be-came fascinated with keys of any kind, and I strug-gled to get them off him once he got his little hands on them. With Pamela, it was my pens and pencils. This worked out well for me because I never lost my keys again and I always knew where to find a pen or pencil whenever I needed one.

Along with keys, Jimmy also loved cars. Before and after every car trip, he jumped into the driver's seat, took hold of the steering wheel, and stood upright on the front seat, play-acting driving the car.

On one bright and beautiful summer afternoon just before I arrived home from work, Jimmy's play-acting turned nearly fatal. After taking the kids shopping, Helen was getting the groceries from the boot with Pamela, when Jimmy jumped into the driver's seat and managed to release the handbrake.

Pamela had somehow managed to break away from Helen's handgrip and walked to the back of the car. She was behind the car when Jimmy released the brake. Helen's old Volvo rolled back and knocked her over before the back wheel rolled right over her stomach.

When I drove up the street on my way home from work, I was shocked to see Helen being driven past me in the front seat of our neighbour's car. She was holding Pamela in her arms, crying and Jimmy was strapped into the back seat. With my heart racing, I turned the car around and gave chase.

The hospital released Pamela the following morning after a barrage of examinations, X-rays and full-body scans. The doctors assured us that she had sustained only minor abdominal bruising as a result of Helen's car reversing over her.

'Happy to inform you both that nothing is showing anything abnormal the X-rays of scans. You have one lucky little girl who has nothing more than bruises from her ordeal. You are fine to take her home as soon as she wakes.' The doctor explained more confused than we were.

Different doctors had different views on why she hadn't sustained serious injury or died, 'The driveway was on a hill and therefore there was less downward pressure on the baby', or 'A baby's bones are more flexible', or 'Perhaps the shopping resulted in less pressure on the baby's side'. We didn't particularly care what had saved her, we were happy she was healthy, and we were able to take her home. It was another miracle, one of many, that never questioned and moved forward with our lives. We were grateful to have one another and grateful were all still together, alive and in one piece.

———

Before we knew it, Peter was christening Pamela back at our Greek church. Peter and his lovely wife Amanda were awesome, they accommodated the priest's demands, and the christening went off perfectly.

Everyone at the christening—family and friends, including George, Debbie and Severino, and both their families. It was magical, with Helen and our two children and my family and my Thal mates and their families. Having them all there was so special. It felt like the starter's gun had gone off and I was beginning my family life with a bang.

26

STORE MANAGER OF THE YEAR, 1989

It was a while before life settled down to a new normal and, by this time, I had been promoted to store manager. Despite everything else going on in my life, in my first year, I was awarded store manager of the year. This was unexpected, especially after being emphatically told by my area manager that I was never going to be a manager and that my best course of action would be to resign.

Upon returning to work after Jimmy's birth, I had been quickly relocated to another store. With the hostility I was receiving from everyone, all senior staff, the other assistant managers and especially the loss of confidence of my manager, I didn't argue. But my new store was still under the same area manager and, when I turned up on my first day, I braced myself for the worst.

My new place of work was in a brand-new building only a twenty-minute drive from home. As I drove in on my first day, I was excited about the new challenge ahead. However, my excitement turned into apprehension when I noticed the area manager walk out with the store manager as I walked in on my first day. They both ignored me, and I headed for the staff room, trying hard to not let their unprofessional behaviour bother me.

After seeing the area manager off, my new store manager tracked me down and demanded a meeting with me to give me a piece of his mind. 'You are never going to become a store manager while you work for me, and it would best for you to leave now and avoid any further humiliation,' he said before walking away.

This sounds familiar, I thought. I knew then I had no chance of career progression there. So I listened to him without saying a word and, when he finished, I simply got up and started my afternoon shift, after a brief meet and greet with the afternoon shift team.

The store was new and desperately needed an experienced assistant manager, as all the staff and the management team were new as well. Durham, the store manager, was the company's top gun. The tall young man of Fijian Indian background had proven himself with remarkable results over the years. He was being groomed for greatness. From day one, his mind was closed to ever building a relationship with me, but he was kind enough to keep out of my way whenever I started my shift. It was bad enough dealing with him, but I also had to deal with all the other assistant manager upstarts, all fighting for the store manager's favour.

Although I loved working at this new store, it wasn't an easy place to work. I was constantly belittled by the store manager and my colleagues. Also, the area manager went to great lengths to avoid me every time he visited. I definitely didn't have any friends at work. In my breaks, I often grabbed the company's policy and procedures, and I read them thoroughly from start to finish. It was to keep myself occupied and out of the way of the bullshit mistreatment I was getting from everyone.

Then one morning shift a few months later, without any warning or preparation, my store manager demanded I stop work and take a two-hour written examination. After voicing my protest, he just smiled and wanted me to follow him without any further hesitations.

Unperturbed, I followed him to his office where I was to take the test under his supervision. Over the next two hours, I completed the test as he sat there with me doing paperwork and ringing around an assortment of places, all while keeping an eye on me. Once I had finished, he took my test papers, placed them into a large yellow envelope and sealed it. Then, without even wishing me luck, he demanded I go back out and finish off my shift.

The annoying assistant manager upstarts were worse than the store manager. They laughed at me and told me I had no chance and that they were going to beat me hands down. These assistant managers were just out of training at head office and felt they had the upper hand because they still had all their training manuals at home.

What was even more hurtful was that they slipped out the reason behind the apparently impromptu test. There were two store manager positions available in our area and the assistant managers with the highest test scores were to receive promotions. All the assistant managers had been told this except for me, allowing the others to study and prepare for the examination. I was furious but I couldn't do a thing.

Three weeks later, the store manager intercepted me before the start of my afternoon shift and asked me to follow him to the foyer of the restaurant. The area manager was waiting there for us.

It's all over, I thought, *they are finally going to fire me.*

'Congratulations Arthur, you had received the highest mark ever in the history of the company in the recent test you took.' The area manager said calmly with a smirk right across his face looking up at me from his chair.

I sat in between them and was astonished to hear the wonderful news.

'Yes, also you have been promoted to store manager at the earliest opportunity. Congratulations. Bozzy, well done!' the store manager, said shaking my hand as they both padded me on the back with sincere happiness for me.

I couldn't believe it. The area manager and my new store manager, who had both told me I would never become a store manager, were now congratulating me on achieving a remarkable ninety-eight percent score in the test and promoting me to store manager level. As a result of my history-making mark, they were both forced by head office to promote me, despite what they thought of me.

When we all walked into the back of the store, it was satisfying to see my colleagues' shocked faces after they were informed of my promotion by the store manager. What was even more satisfying was finding out their pathetic test results as none of them had even received a pass mark.

I was on my way now and, from that day onwards, the area manager, my store manager and I became friends. I couldn't believe that these two people who had loathed me so much would end up being two of my closest friends.

Soon I was taking up a position as store manager in a small store a few suburbs away, which I enjoyed very much. I always had a knack of taking charge by leading by example and avoiding being a tyrant. 'Show, tell, do and review' was my motto and everyone knew it.

We all worked hard once I took control and, after hearing about all the secret store checks earmarked to be undertaken in the area by an assortment of mystery shoppers, I got ready for our store to be the best of the best. Ratings and feedback provided by mystery shoppers were used to rank the stores and determine the best stores in the area, as well as highlight those with aspects that needed improvement. The manager of the winning store received the prestigious store manager of the year trophy and the choice of managing any store in the area.

With only nine months left until the end of the year, we knew we had no chance of winning store of the year. However, my team and I weren't discouraged. If we couldn't get twelve out of twelve top scores, then we could dig in and aim for the maximum of nine out of twelve. In the meanwhile, I had something to look forward to as the company had selected Fiji as the location of the manager's conference that year.

The next thing I knew, I was in gorgeous Fiji at my first managers' conference, sipping on amazing cocktails and washing them down with an assortment of cold beers. The attendees were divided up into our areas, sitting on cushions around tables loaded with an abundance of food and drinks and a huge array of Indigenous Fijian decorations.

The entertainment included music, dancers, and a spectacular fireworks display. It was sensational and memorable. Then there were a few speeches and award presentations and, by the time they presented the last award of the night, store manager of the year, I was totally drunk.

I was so tanked up, I imagined I heard them calling out my name as winning store manager of the year. They announced that my store had been rated number one by their secret mystery shoppers for nine consecutive months and that was enough to win the trophy.

'Arthur, Arthur, they are calling you! Get up, you won, they are calling you!' I heard from all around me, but it wasn't registering.

As I reached out to grab another drink, I noticed everyone looking at me and yelling, but I had no idea why. I was still sitting there, dumbfounded, when my area manager came over to congratulate me. He dragged me to my feet and, after pointing me in the right direction, gave me a slight shove. Now my only problem was making it to the stage.

I have always been a light drinker—two beers were my maximum—but on this occasion, feeling hot and thirsty in tropical Fiji, I must have had three beers and God knows how many cocktails. I can't remember exactly how much I drank but when I woke up in my room the next day, I couldn't find the trophy and thought it was all just a dream.

To my surprise, at breakfast, I saw another manager walking around the breakfast buffet with the trophy I thought I had won. After walking over to congratulate him, I was bombarded by well-wishers congratulating me. The store manager who was still intoxicated holding the trophy handed it over to me with tears in his eyes and he said, 'I wish I had won this beautifully trophy!'

It was true, I had won store manager of the year! There was no doubt about it, the trophy had my name clearly displayed on it. I took it off him and, after shaking everyone's hands, I tried to have breakfast between all the interruptions. It was amazing, with only nine months of the year left after I was promoted to store manager, I had assumed I had no chance of winning.

I waited until lunchtime when I was completely sober, to claim my prize. I approached my area manager and told him I wanted to manage the Mount Druitt store. Because it was the biggest store in the country at the time, I knew this request was not going to sit well with him. Prepared for his initial rejection, I began my demands lightly before stepping up to an all-out assault. Yes, he rejected my choice at first and the more he asked me to select another store, the more I pushed my argument for Mount Druitt. Refusing to give up, I finally wore him down the second week after our return from Fiji.

So, in my second year as store manager, I was appointed manager of the highest volume store in the country. I was ecstatic. I had wanted the freedom to be creative and innovative without first having to run my ideas past imprudent and mediocre individuals, and now that freedom was mine. But my thirst for greatness was not yet quenched. I was at the helm of a huge ship, ready to steer it to great places, my compass was pointing in the right direction, and I didn't look back.

Securing the position of store manager at Mount Druitt was important to me on a personal level. It was the first store the company had placed me in during my six weeks of training before I was assessed and allocated to a permanent position in another store after qualifying as an assistant manager. When I'd started my initial training at Mount

Druitt, my confidence had been low after being told I hadn't done well in the general knowledge entrance exam into the company. But they took me on anyway because they understood I was working seven days a week in my family's takeaway.

Furthermore, I was nearly terminated in the middle of my training at the Mount Druitt store as a result of opening the back door without the store manager present. She had been running late one morning and I'd decided it would be a good idea for me to go in and get started so we wouldn't be rushed off our feet preparing for opening time. I was wrong. It wasn't such a good idea. As soon as she arrived, she called the area manager and demanded he fire me. I was grateful he was feeling benevolent that day and chose not to listen to her. Things never got better between the store manager and me after that, but I was thankful she passed me at the end of my training, if only barely.

So it was from that early stage of working for the company that I had wanted to be the store manager at Mount Druitt. I wanted so badly to show that manager how much she had underestimated me. On that day, while feeling vulnerable and disheartened, I had made a promise to myself that I would be store manager of the Mount Druitt store. Yes, I never imagined fulfilling that promise so quickly. It was a dream come true.

It took me about six months after taking up the position as store manager of Mount Druitt to feel completely comfortable. By then, I was ready to take it up a notch or two to motivate the team. I had inherited a huge team who didn't like my style of management. They loved being left alone and I liked being in constant touch with everyone. I needed everything done my way but without micro-managing them.

I had constant daily battles, and no one would give me a break. So I decided to do something about it before it got worse. I knew of a new independent movie cinema nearby and I gave them a visit, in the hope we could do a deal for our mutual benefit. To my surprise, I struck a deal with the director of the cinema. I agreed to display details of movies currently showing in his cinema in the foyer of my store in return for free pre-movie advertising of my store on all four of his cinema screens. Furthermore, he supplied me with ten free movie tickets every week.

Immediately, I put these free movie tickets to good use as weekly staff prizes to motivate my team, which finally got me the attention that I needed from everyone, especially from my frontline staff. However, my plan worked so well that, before I knew it, my store's soaring figures had created such a huge commotion across the company, I had head office breathing down my neck to see what changes I had implemented to achieve those results.

'Arthur, who approached who first?' My area manager asked me in the foyer of my store as the suits from head office all eagerly waited for my reply.

'I first approached the cinema manager at his office and he like what I had to offer.' I said as I looked at each and every suit, one at a time, in a slow and confident manner.

'What did you offer them?' The lead suit asked as they all moved in closer.

'I humbly asked him what he could do for me, if I would allow him to place his what's showing now posters and free-standing cut out displays in my store for free!' I replied with a smile.

'Well, what did he say?' The lead suit asked now even more curious.

'He told me he could give me free movie tickets every week also advertise my store in all his four cinemas, for free. Plus, he wouldn't charge me to make up the film for the ads on all four projectors. I agreed and we both shook hands on the deal. I used the free movie tickets to motivate my staff every week and the rest is history.' I explained and then they all ambushed me with a barrage of handshakes.

They wanted to know about everything. For the next few weeks, it was not unusual to have a dozen or more suits in my store daily, talking to me and all my staff. They even got me to take them all over to meet the cinema director and get him to show them the free advertisements that were being shown just before the start of each movie.

Satisfied that my reciprocal arrangements were free of any costs to my store, they finally got out of my way, and I went back to inspiring my team. Later that year at the store manager's conference in Sydney, I was surprised to be presented with the Local Store Marketing award of the year. It was an unexpected and wonderful achievement in my first year of managing a monster of a store.

The following year, I took on the role of acting area manager. Not long after that, I was promoted to training restaurant manager, responsible for selecting and training store managers for the entire NSW state.

By the time I left the company in late 1993, I was proud of my achievements at KFC. From when I'd started in 1986, all I'd ever wanted to be was a store manager. KFC was an amazing company to work for and my ride to incredible heights over a short period was remarkable under my circumstances, but I soon learned that my determination and desire wasn't enough.

Secretly, my success was undoubtedly a result of Helen's continuing support, lifting me from my deepest depths of despair to the highest of highs. She was wonderful with her wise words and thoughtful reassurances. With Helen at my side, I felt as if I could take on anything.

I would soon discover that my health was spiralling downwards fast, and I would need Helen more than ever before.

27

INSERTION OF A PORTACATH, 1991

In early 1988, The Royal Alexandra Children Hospital at Camperdown finally transferred Peter, George, and me to the nearby adult hospital at the Royal Prince Alfred (RPA) for ongoing treatment, and Severino to The Prince of Wales Hospital at Randwick. The children's hospital had never had Thal patients live as long as we had and didn't know what to do with us.

We were breaking new ground. We were the oldest Thals ever and it was hard for the doctors and hospital administrators to decide whether to relocate our treatment to another hospital or maintain the status quo and wait a little longer until we succumbed to our condition, like all the Thals before us.

By then, Peter, George and I were all married with children or had children on the way. Peter and his wife Amanda were expecting Michael. George and Debbie had welcomed Christopher the previous year, a few months before Jimmy had been born. Severino was still a bachelor, and he loved the isolation his new hospital provided. Occasionally, we would catch up with Sev, but it was mostly the three of us in each other's pockets and routinely going to the RPA for our monthly transfusions.

On one particular monthly blood transfusion day in February 1991, not long after I turned thirty, I received awful news. I was told I was to be prepared for a portacath insertion before receiving my treatment. Flabbergasted, I demanded to see the doctor who had ordered this before anyone touched me.

Once they'd tracked down the day stay doctor on duty, he explained that my veins were no longer suitable to access for transfusions, as a result of damage sustained due to the repeated IV needle insertions over the years.

'I'm sorry, I'm not willing to insert your IV, you need to see the doctor.' The senior head nurse bellowed out for all to hear her.

'I just walked in, what are you talking about?' I said before taking a seat.

That, combined with my negligence in not keeping up with my Desferal pump treatment at home, had forced them to make this crucial decision.

I soon found out that since we had recently been shuffled over to the day stay unit for our treatment, they had been recording our ferritin results from our monthly cross match blood tests. I knew the cross match was needed for haemoglobin results to determine how much blood we were all required to have transfused the following day, but I hadn't known what the ferritin was for, and nor did I care.

I knew the normal range for haemoglobin for men is between 13.5 and 17.5 grams per decilitre and for women between 12.0 and 15.5 grams per decilitre. My haemoglobin level each month was averaging between nine to ten grams per decilitre, which meant I needed three units of A Negative blood every month. If it was higher, I needed less blood but if it was lower, of course, I needed more blood.

The blood was now packed in plastic bags with an average volume of about 200 to 300 millilitres of packed triple-washed red cells per bag. The purpose of washing the red blood cells is to remove plasma which contains substances that may trigger an antigen-antibody reaction. It was important for us to avoid having adverse reactions to our regular blood transfusions.

I soon found out that Ferritin is a protein that stores and releases iron in the body. Ferritin levels indicate the amount of iron stored in the body. When the doctor told me that the normal ferritin level for men ranged from twelve to 300 nanograms per millilitres of blood, and half that for women, I didn't know where he was going with his explanation. But the earth fell away beneath my feet once he explained that, due to my non-compliance in keeping up with my Desferal pump treatment at home, my ferritin level had increased to a critical and never-before-seen level of just over 30,000. In comparison, both George and Peter's ferritin levels were hovering under the 3,000 mark. I was ten times higher. My soaring ferritin level, and therefore extreme iron overload, had forced them to elect for the insertion of a portacath as soon as possible.

'What do you mean, you don't want to focus on the past. We are here because I'm too worried about the constant dramas going on here with all you lot to think about anything else. What is it this month, because up to now, it's been my veins?' I said in total frustration not knowing what to expect.

When the specialist, doctors and head nurse explained that a portacath has a reservoir compartment that has a silicone bubble for needle insertion with an attached tube called a catheter, I wasn't interested. But I panicked once I heard that the entire thing runs from a portal which is surgically inserted into a vein, usually a huge artery called the superior vena cava, and their preference was to insert it in my chest. When they started telling me about all the benefits of never again getting prodded and probed by sharp needles while a doctor or nurse tried to find a vein, I lost it. I didn't want to hear it and made it clear that I was in no way interested in having a portacath inserted. I demanded another option.

'Wow, I don't think so, no way. I know what they are but sorry, I'm not interested.' I reply fell on deaf ears even after repeating it a few times.

Unfortunately for me, the consensus from them was to insert a portacath and I was put in a difficult situation. Although I knew of many Thals half my age who had successfully had a portacath inserted and made it work for them, I was still against it. But they insisted a portacath would be necessary to run the Desferal pump twenty-four-seven in an attempt to lower my ridiculously high, near-fatal, ferritin levels. They announced that if my ferritin levels stayed that high or increased, then my death was imminent.

With a target ferritin level of under 1,000, I knew it would be near impossible to get it down to that safe level without the portacath. I also knew that even if I fully complied with my Desferal treatment, I may not survive long enough to even get halfway to that target. There is no doubt that, at this horrifying stage of my life, I was literally the ultimate iron boy.

The doctors gave me no guarantees that I would not succumb to iron overload in the meantime, but with my Desferal pump running twenty-four-seven, via a portacath, they hoped to reduce my ferritin level to a safe level as soon as possible. They made it clear that it was now simply negligent to run my Desferal pump for only ten hours a night for four to five nights a week with the unrealistic expectation of reducing my iron load to an acceptable level.

I knew I had stuffed up over the years by not complying with my Desferal treatment, but they'd told me when I'd started using the pump that the damage has already been done and I could still die at any time. Now at thirty, I couldn't believe I was still alive. I wondered if I had used my pump consistently all this time, could I have avoided ending up in this position? I was confused, but one thing I was clear—I wasn't going to get a portacath inserted.

Before I knew it, I was surrounded by the entire blood clinic staff all insisting I agree to having a portacath inserted immediately to run my Desferal pump twenty-four-seven if I wanted any chance of staying alive. This decision was put forward so quickly and in such a well-rehearsed way by all the doctors, specialists and nursing staff that I knew their approach had been coordinated in advance. Although it was an attractive option and the best approach for me to rapidly lower the high iron levels in my body, while giving my veins a well-earned rest, to the doctors' dismay, I dismissed their recommendation.

At the time, I had a phobia of portacaths and the thought of anyone inserting one into my chest or anywhere else in my body horrified me. Although I found myself not completely ruling out the idea, I was determined not to agree to a portacath until I had a good go at increasing my Desferal treatment compliance back to 100%, as it should have been in the first place.

With two children under four years old and my wife in the process of training to go back to work, I kicked back by standing up for myself at a time I needed to the most. This enabled me to temporarily put my emotions to one side so I could focus on what I needed to do to get myself out of a dreadful and terrifying situation.

But, on the worst morning of my life up to that point, how was I going to convince these determined doctors, specialists, and nursing staff to give me another chance to turn things around? Not only was I was feeling physically weak, but I was also traumatised by this new information I'd had no time to absorb, and I was confronted by all of these medical professionals who insisted I allow them to surgically insert a dreaded portacath into my chest.

Before dealing with my emotions, I needed to deal with all the new information and thoughts and fears racing round and round in my mind.

The doctors tell me I have critically high and potentially deadly levels of iron stored in my body. What's more, they say it is my fault. They say it's a direct result of my noncompliance in self-administering Desferal, the so-called miracle treatment that has kept me alive much longer than I ever expected to live. But now, my body is overloaded with iron. My organs could fail at any time. My death is imminent.

What's more, my poor over-used and abused veins are giving up on me, they are no longer suitable for inserting IVs. My doctors are refusing to put up my IV, even for my regular blood transfusion. The doctors, nurses and specialists are insisting I have a portacath surgically inserted to run Desferal twenty-four-seven, and also to run my regular monthly blood transfusions. Without Desferal running twenty-four-

seven, I'm going to die, they say, but then again I might already be too far gone. Will I die anyway?

I'm not going to see my family again. This really could be it. I could die here. Have I finally reached my expiry date? I feel weak and light-headed. I need my regular blood transfusion now! My emotions are getting out of control. If I survive, will I be able to keep on working? Will I lose my job? How can I provide for my family? The very thought of the bank taking our house and my family being left homeless overwhelms me and leaves me sick with fear.

I feel as if I have been ganged up on and ambushed by all the doctors, nurses, and specialists. I'm feeling outnumbered and alone. Why didn't they inform me or include me regarding my health progress? They have been monitoring my ferritin levels for months. How could things have gotten so bad without me knowing? Why didn't they tell me what was going on? I wasn't prepared for this.

I'm confused. I don't know what to do. But the thought of having a portacath inserted into my chest petrifies me. Sure, I concede that there is some merit to the idea, but I could only agree to it as a last resort. Then again, if my ferritin levels are so dangerously high and my death is imminent, as they say it is, then should I consider this to be a last resort? What if I could get my ferritin levels back under control *without the portacath? I'd like to at least try. But what if I die trying?*

I wish I could talk to someone about this, someone who would listen to me and not just try to push their agenda. I'd like an unbiased second opinion. I feel as if it is the medical professionals against me. I wish someone was on my side. They are getting tired of me not wanting to toe the line and agree to everything they suggest. Now they are insisting. Yes, they are getting more demanding and impatient, and some are even openly hostile. I'm struggling to handle their bad attitudes along with everything else going on today.

None of them is interested in listening to what I have to say. They don't care about me or what I want. I'm just another number to them. Yes, they are the experts, but this is my body and my life, so surely I should have some say in my treatment. I feel like I have no choice and no voice, as if my life is out of my control. I need to rein in my scattered thoughts and emotions.

With a strong authoritative voice and a determined manner, I took control of the situation. I finally convinced them to at least temporarily delay the portacath option. They also reluctantly agreed to get a more experienced doctor to insert my IV for my blood transfusion that morning and to organise a more permanent alternative for my future blood days. It was all over in seconds but not without them also agreeing to leave me alone to try and increase my Desferal compliance back to 100 percent.

After a lifetime of caring and individual attention at the children's hospital, since moving to the RPA each of the Thal patients found ourselves just another number in the system—we became almost invisible. In those days, the day stay nurses and doctors were all too busy and seldom considered consulting us or including us in our treatment plans, and when they did, it was tokenistic. Even worse, the day stay unit could be a downright hostile environment, staffed by some unprofessional and uncaring workers. On one occasion we were told, 'Stop calling for nurses to change your blood packets. Can't you see we are all busy? You aren't even sick. We have sick people in here who need our attention!'

None of us said a word in response, even Peter controlled himself. We simply put up with their heartless treatment. We were in unfamiliar territory. They were so unprofessional in those years.

After I stood up for myself that morning, annoyed I was getting my IV put in elsewhere, the day stay unit staff all made a huge fuss about my veins being so bad that they all refused to put a needle in me if my specialist wasn't available. Yes, we hated the day stay unit days.

Dr Harry Kronenberg was the head of Haematology at the RPA at the time and he and his wonderful team had looked after us initially. He'd made the transition from the children's hospital seamless by keeping us all together and ensuring we had professional people looking after us in the early days. So I approached Dr Kronenberg and told him about our situation in the day stay unit and how we were being treated by the staff. Soon after that, things changed for us but not much.

From then on, I was permanently referred to a specialist doctor who was responsible for putting up my blood IV on every blood transfusion day, even though we were still at the day stay unit. Also, we immediately noticed staff attitudes change towards us right up until we were relocated to a specialised unit that treated only Thals and Haemophiliac patients.

The person in charge of this newly created and wonderful blood clinic was Fiona Rennison. Her attitude and professionalism quickly gave me the edge I needed to let go of all the negativity I had built up over the years. No longer did I need to worry about my veins as each month at the blood clinic, Fiona would always find my vein; she never missed.

At home, even though we had two children under five, Helen helped me get back on track in using my Desferal pump no fewer than five nights per week. I had wasted enough time and now my head was clear to focus on reducing my ferritin level. Six months later, my levels already showed huge improvement, but I knew that it probably wasn't enough to avert the inevitable and I braced myself for the worst.

Death touched our lives shortly afterwards, but it wasn't what I was expecting. In early in 1993, my best man from my wedding and Jimmy's godfather (Koumbaro), Andrew Patsos, died from leukaemia. This shook my very foundations, and I was paralysed with fear for quite a while because I was still dealing with my mortality. But what further baffled me was that I couldn't understand how any illness could take down a man built like an ox. He was Hercules, not only to his closest family and friends, but to anyone who met him.

I don't understand; how can this happen so fast to him. No not him, it just can't be, it should be me. I asked myself over and over outside Andrew's hospital room waiting to go in.

Both Helen and I already had on our personal protective equipment by the time the door opened but it was useless in protecting us from the doctors' briefing, the families' enormous anguish, or preparing us for Koumbaro's catastrophic condition we found him in.

We said what needed to be said while sobbing uncontrollably but he would have none of that. To our surprise we found ourselves being comforted by him and it was both calming and beautiful. But it wasn't long until I realised our Koumbaro was still mentoring me.

I lost my best man that horrible day in 1993, but I was so grateful to have had him in my life. I gained such incredible strength and wisdom from Andrew at a time I'd needed it most.

Less than three years later, tragedy struck Andrew's family again with the loss of Andrew and Christina's beautiful son, little Con. The baby of the family had been growing into someone like his father and everyone was in awe of this likeness. We couldn't believe something so tragic and heartbreaking had happened. However, amid their heartbreak, Andrew's wife Christina, and their two gorgeous daughters Vicki and Joanne did something I would never forget. They started to live again, knowing that little Con was with his father.

I knew Andrew and his beautiful son would have been proud of the three of them for picking themselves up from the deepest, darkest depths of despair and live their lives to their fullest. It was extraordinary to see. These were fearless women who gave me the strength to do the same. They were living for their loved ones. I started thinking of doing the same. I had always been afraid to start new things on my own but now I found some courage to really start to live.

Not long after, the girls both got married to two wonderful young men a few years apart from each other. Joanne and her husband Andrew became proud parents to a beautiful baby boy and a gorgeous baby girl. It wasn't too long that Vicki and her husband Peter were proud parents as well with two gorgeous baby girls and a beautiful baby boy. I'm sure Vicki's local government aspirations even shocked her amazing husband Peter, who by then had racked up over twenty years in state politics by the time she ran for local councillor.

Koumbari were both so proud of Vicki when she became school councillor in her sixth-grade class all those years ago. Koumbaro would have been absolutely even more proud of her going into local government after starting a family. She successfully got voted in as councillor for her local government area and she services her local community steadfast and with distinction on all local issues. But Koumbara once told to me that Koumbaro would have been most proud with their girls married to remarkable men who are even more remarkable parents themselves.

Even these days, you still wouldn't find Koum-bara Christina missing a weekend cleaning up and placing fresh flowers for her two loved and very much missed men. The family are all in wonder with Christina how she works so hard also con-stantly supporting the elderly, which she still loves doing, juggles her gorgeous grandchildren while never letting down in keeping her two favourite men updated flawlessly week after week.

28

BUSINESS OF MY OWN, 1994

Once the decision had been made to go in business with Helen's brother Nick, I found myself neglecting my health again. Although this also meant I was jeopardising my family's best interests, I didn't give it much thought and proceeded without listening to anyone. I had an opportunity in front of me and I was determined to see it through.

With my business plan in hand, we each invested both our money to build a chicken delivery business. We leased vacant premises which had once been a hardware store and outfitted it with huge cool rooms, cookers, ovens, and a charcoal rotisserie. A counter was set in place, the foyer furnished, and we purchased the latest computers that meticulously organised our home delivery systems.

After the debacle of working with my brothers in our family business, I wanted a business of my own so I could prove to them I had what it took to succeed. I designed and developed the first chicken, pasta, and salad home delivery business in the western suburbs area, and I arrogantly and ignorantly wasn't going to let anything stop me, not even my health. It wasn't bad enough that I put our money on the line, but Helen's brother invested his money as well. I was putting the relationship between our families on the line if things didn't work out.

Within six months of opening, we had sales exceeding our expectations and in one particular month, average weekly sales were remarkable. We were proud of our efforts. But, at the six-month point, the wheels fell off and the situation declined rapidly. Our quarterly electricity bill was averaging $10K and we were just keeping our heads above water without this extra expense.

We had outfitted the shop well and did have enough of a buffer saved to take this huge hit, a few times by now. But I needed another six months to get the place firing and increase our sales further, but when I saw the anxiety on Nick's face, I knew I had a bigger problem to deal with.

No matter how I approached him, I couldn't convince him that we needed to focus on sales to weather the following six months of overheads. Finally, to avoid further upsetting him, I reluctantly agreed to his inane cost-cutting plans. At a time when we should have continued our aggressive letterbox advertising and ensured our full menu was always available, putting a halt on all of this to save money buried us. I didn't complain—I was too sick to argue with him after he made it clear I was to listen and give him the same courtesy that he had given me in the beginning.

We were doomed, and I just held on and took my punishment. It didn't take long after that decision and it was all over by the tenth month. I knew I deserved it for getting him involved, as he didn't understand, and I didn't hold any animosity towards him regarding his decision. But his hurt soon turned into anger and, by the time we closed the doors of the business, our friendship sadly also ended, at least for a while.

With my ferritin levels now back up to a critical high, I found myself out of work and only weeks away from having our house repossessed by our bank due to three months of defaulting on our mortgage repayments. I had let everyone down, but I had let myself down the most. If I had looked after myself, I would have safeguarded my family and also not put Nick in over his head in something he knew nothing about.

We had been living in Minchinbury since 1990. Taking advantage of house value increases as a result of new freeway access, we had sold our house in Pasadena Place for three times of what it first cost us. This was great considering we only had it for less than five years. We had found a gorgeous six-month-old house in Minchinbury and quickly snapped it up. It was huge, with four bedrooms and a double garage. Although it was in the same street as Helen's parents' house, when we looked inside we fell in love with it.

Now we were on the verge of losing this beautiful home, and I was responsible. I needed a job, and quickly, to avoid that happening. When I discovered a McDonald's franchisee in a nearby suburb advertising for managers, I applied immediately. What I knew of McDonald's unique management training program made the job appealing and I felt I was well-qualified for the position after eight years of experience in fast food.

A few days after I submitted my resume, the store manager called me in for an interview with the owner and store manager, and I started working there the following week. Both the owner and the store manager were impressive individuals. I liked them both right from the start. Eleven assistant managers plus the store manager made it a management team of twelve.

My first shift started at eleven p.m. on a Friday night and, after turning up an hour early, I was flabbergasted to see how busy it was. I couldn't even find my tie by midnight. It was a baptism of fire for me—I had never experienced such a volume of customers and this was at midnight on a Friday night. It was madness of the likes I had never seen before. And it was like that all the time, day or night.

On one particular graveyard shift during a weekday that first week, I had The Honourable Gough Whitlam staring down at me, waiting to order, as I was left manning the fort all alone on my first of many nightmare shifts from hell. He had appeared out of nowhere, with all his stature, desperate for something to eat, coming from a function that he hadn't wanted to leave. Unfortunately, I had just sent the entire staff on a well-deserved break, and I was paralysed with fear over what to do next after being confronted by him at the front counter.

I told him I was new and how impressed I was with both the front counter and kitchen staff and had uncharacteristically sent them all on a break together, after a busy and hectic night. Still stunned by his appearance, my honesty worked and, considering how hungry he was, he just laughed when I told him, 'Sorry, but you have next to no chance to get something to eat from me!'

It must have been a little before midnight when

he'd first walked in, and we both waited almost twenty-five minutes for the staff to return to take and cook his order. They were speechless once they spotted us both leaning on the front counter drinking our coffees, deep in conversation. He opened up as a result of my lack of proficiency with the registers and cooking equipment. What a unique and exceptional individual with a remarkable insight into life's challenges that was a gift from God. Sharing those twenty-five minutes was a gift he gave me, and I never forgot what he said about how adversity does not build character, it only reveals it.

The store was open twenty-four-seven and, in those rare periods with no customers, which only lasted for very short periods, if we didn't spend the time restocking and preparing for the next influx of customers, we were done for. Trying to catch up when it was busy was futile—we would blow out waiting times if we tried to stock up amid the mayhem.

Right from the beginning, the management team treated me with hostility, which became all the more noticeable after I took up the position of second in charge. In some ways, their hostility was justified because I didn't know anything about McDonald's, yet I had been promoted to second in charge of the store. Most of them had worked hard for years and hated that I had come in and got the position over the top of them.

If that wasn't bad enough, the store manager, who was the oldest of the staff at barely twenty-three, asked me, 'What are you doing, at your age, working in McDonald's?' At only thirty-four, apparently I was too old.

I had enough of my own worries, so I didn't bother engaging with any of their issues. Instead, I just focused on my training and responsibilities. Now that I had a weekly income, I wasn't about to jeopardise it by tangling with my co-workers. My remuneration package may have been acceptable for a twenty-year-old, but it was far from acceptable to me. However, I was thankful to have a job.

Within my first month, the owner agreed to allow me to enrol in the first of the five management qualifications to become a store manager. My resolve to not argue over pay and avoid conflicts with my colleagues, as well as surviving the ridiculous number of customers, had paid off. The renowned management training I had heard so much about over the years was now being offered to me. I was delighted and I grabbed hold of the opportunity with both hands.

I couldn't believe my luck that the store owner was going to spend the money to put me through this first unit of manager training at the famous Mc-

Donald's Hamburger University at Thornleigh Head Office.

Six months later, in an unprecedented time-frame, I had successfully completed four of the five required management training programs. Although designed to be completed over a four to five-year period, I stunned everyone and annoyed most of the trainers at head office by completing the four programs in such a short period. If not for their resistance, I would have completed the fifth one, but they refused to accept me back. I didn't know what the problem was because the franchisee was paying for my training, and I was passing them all.

In hindsight, it was fortunate I wasn't able to do the last program because not long after that I decided to leave McDonald's and was glad I hadn't racked up further expenses for this wonderful store owner. Even though I loved working at McDonald's, I had found a new job that paid not much more but it was working normal weekday hours, so for the sake of my health and young family I had no choice but to tender my resignation.

My position at McDonald's did keep the bank at bay, but it wasn't paying all the bills. Working harder hadn't resulted in a pay increase, nor had more hands on at the restaurant or gaining qualifications. After watching Helen go back to work to help pay the bills, I needed to look elsewhere to work normal hours, so I can do my pump, improve my health and be there for Helen and the kids.

At the time I left, there were lots of disputes between the management team and the owner. It was sad what everyone put this wonderful man through. I didn't have the heart to tell him I was leaving, so he probably thought I was siding with the management team and had resigned because of that. However, it was simply because I needed to start regularly using my pump again and better supporting my family.

My new job was as an employment consultant for a disability organisation. At the restaurant, we had always been hiring people with disabilities to do an assortment of jobs; many had foyer cleaning duties and others worked in the kitchen. It was while on duty at McDonald's that I'd met the manager of the disability organisation, responsible for selecting the right people for the various jobs we had available at the restaurant. Sometime later, this wonderful person told me about the job vacancies at her workplace and I decided to look into them further.

Before I knew it, and with her help, I was an employment consultant helping people with a disability find open employment in award wage jobs. I soon realised the job was nine to five and I started to love these business hours. For the first time in my working career, I wasn't working nights or week-ends and I was back using my pump regularly in no time.

I was so grateful to the manager who had helped me get the job. Without her, I wouldn't be getting paid to help these beautiful people who were our clients. The more I worked with them to get them a job of their choice, the more I could re-late to them. It was as if I was helping someone like me to find a job. I couldn't believe it, I had so many things in common with them, they were no different from me, but they didn't know it.

Soon I found I was working hard for them be-cause I would have loved to have had someone dedi-cated to helping me find a job. In my first year after my induction training, I was only required to find jobs for three people, but in the nine months I worked there, I was successful in placing nine people with a disability in open employment paying award wages. Although not all of them stayed in their new jobs, I enjoyed securing all nine positions.

While doing the job, the highs were really high, but the lows were surprisingly low. I wasn't ex-pecting the ups and downs. However, the CEO of the organisation, was constantly hard-pressed deliv-ering vital staff training so we could navigate the highs and lows of the job.

When the sad day arrived for me to tell him I

was leaving after only nine months, he was amazing yet again. In fact, after explaining to him I had been offered a manager's position in a disability organisation called Self Advocacy Sydney, he looked over my job description for me and encouraged me to take it on if I wanted it.

From the start, I had found him a remarkable man for many reasons, but this took my breath away. He was wise with his words and everyone in the organisation felt his drive and passion when it came to getting jobs for these special individuals. Every time I heard him speak of how important everyone's first job after leaving school is for them, I knew he too was reflecting inwards like me. I didn't know the man all that well then, but I thought he was a special individual, and I was fortunate to work for him.

When the beautiful person who had helped me get the employment consultant position supported me in my decision to accept the new job, I found it much easier to make the move. She has remained an inspiration, a rare individual I have been fortunate to know. Once I had her support as well, I was off to rise to even greater heights than ever before.

29

SELF-ADVOCACY SYDNEY, 1996

After turning up fifteen minutes early for my job interview, I backtracked and raced out of the building, horrified upon seeing all the weird and abnormal faces. I don't mean abnormal like pimple scaring or other face imperfections, I mean profound birth defects.

I knew I had the correct address for Self Advocacy Sydney because I had been told the building was originally a residence that had been refitted into multiple offices and the backyard was their carpark. I don't know what I was expecting and nor do I know why I reacted the way I did. After all, I had been working with people with disabilities for some time, so I didn't know why I was so taken aback.

On my way back to my car, I abruptly stopped and decided to turn back and give it another go. Once back inside, a lovely looking and well-dressed lady pleasantly surprised me and after a quick introduction escorted me back into the house.

As I quietly took a seat in the foyer and looked around at the same faces I had glimpsed before, I realised there was no escaping now. The longer I sat there, the less notice I took of the physical appearances of the people around me and was then able to see their individual personalities.

By the time I was called to the interview room, I had adjusted to only seeing people, not their disability. They got me to sit in front of an interview panel of five, who took turns asking me questions. Thirty minutes later, I was asked to wait in the next room for the final stage of the interview. In the adjacent room, I found four other interviewees, and no one was talking to or even looking at each other.

What the hell am I doing here? I wondered. *I have no qualifications or experience in this field.* But I didn't want to leave either, so I just went with it. The ad for the position had stated their main requirement was experience in working with people with a disability.

Self Advocacy Sydney Incorporated (SAS) is a disability organisation that is run by, and for, people with intellectual disabilities that supports people with intellectual disabilities to achieve individual self-advocacy outcomes. Their board of management was looking for a manager who could help SAS achieve federal government certification for recurrent funding.

Although their board consisted entirely of people with an intellectual disability, the interview panel included three people with an intellectual disability and two independent people supporting them. For the last stage of the interview process, they gave the five of us forty-five minutes to put together a one-page PowerPoint presentation on any topic of our choice to present to the panel.

This was the first time I met Robert Strike. One of the co-founders of SAS, Robert is a remarkable individual who spent his entire childhood growing up in a ghastly state-run disability institute. When he finally got out at seventeen, he wanted to help others like himself and ended up starting SAS. After listening to his remarkable story of being one of the co-founders of SAS and doing all this with an intellectual disability, I started feeling as if I was in the right place.

Each of the interviewees was allocated a computer, and the panel members stood back and gave us space so we could get on with the task. I was put in front of a doo-stop of a laptop so old, it was black and white. The topic I chose was ice breakers ideal for all meetings. I put together the best presentation I could, given the antiquated technology, without complaining, unlike the other candidates. The others bitched and complained so much that they ran out of time and didn't finish their presentations within the allocated time.

When they displayed my presentation, I panicked a little when I realised my font size was still set to large. On the horrible, ridiculously small laptop they had given me to use, I had purposely increased the font size to display large letters in an attempt to avoid making mistakes and, in my haste to finish on time, I forgot to reduce the font back to a normal size 12 font.

It was my luck that they not only loved my topic, and also the fact that I didn't once complain about my equipment, but what got me the job was the bloody large size 18 font. They loved it, believing I was already attentive to what people with an intellectual disability preferred and demanded for viewing presentations.

When I left the building that day, I was in such a different place from where I had been when I had arrived, the contrast was astronomical. They all made me feel I was the right person for the job, even though I had been hesitant at the beginning.

But, especially after hearing all their wonderful praise and being offering the manager's position, I felt somewhat embarrassed and guilty for being so taken aback by all of them when I'd first arrived. Maybe I was the least qualified and the most unexperienced manager out of the five who had been shortlisted, but I got the job and that's all that counts.

Everything made sense to me soon after I started when I met the board of management, who explained to me they specifically didn't want to hire someone who was already working within the disability sector. They wanted someone to manage the organisation by executing all their decisions. In other words, they wanted someone to listen to them and they wanted all presentations in a format suitable for them to understand.

The board comprised of ten people with an intellectual disability and a couple of support people who were crucial in helping to break down things to assist them all to better understand issues like finances and whatever was on the table. Even though I was still initially taken aback by their physical appearances, I was impressed with them and how competent they were with all matters presented to them.

The introduction of the Disability Services Act

1986 provided a comprehensive framework for the funding and provision of disability support services. SAS was federally funded that year on April 1 to educate, support, train and provide information to people with an intellectual disability regarding all their self-advocacy individual outcomes.

People with an intellectual disability who were living in the community knew what others like them in the community wanted. Their management model was impressive because the organisation was run by, and for, people like them. There was nothing tokenistic about the board members—they were genuine, and all legitimately voted in by the members. It was extraordinary to see.

They were in their tenth year of operation and the funding body gave them less than five years to put together their policies and procedures to gain certification for recurrent funding. Time was running out. What little documentation they had was handwritten in pen on writing pads. There was a lot to do, but we had almost five years to complete it.

The interim manager, Berinda Karp, on secondment from her disability consultant TAFE College position, did her best to show me the ropes on my first day. Although she was looking forward to going back to her fulltime position, I noticed she had a special bond with the team and was going to miss everyone. I envied the relationships Berinda had built with the team, and I knew then I had some big shoes to fill.

Everything was going well, and, by the end of the day, I knew Berinda wanted to say something to me but was waiting for a good time to drop the bomb. Just before we finished up for the day, she finally told me what had been on her mind the entire day.

'Arthur, I need to caution you about a staff member named Julie Strike who's currently on annual leave.' Berinda spurted out with a distress look on her face.

'Who?' I replied looking straight back at her.

'Julie Strike.' Berinda repeated with a half-smile.

'Is she related to Robert?' I answered.

'Yes, that's his wife who's on annual leave.

'What's the problem, I get on with Robert?' I said confused.

'Look, even though I hold Julie with the highest regards, I feel you should keep away from her because Julie doesn't like change or new staff members. I feel Julie wouldn't like working with you due to her not being involved in approving your appointment. You see, if she not involved with the entire recruitment process, then she will make it her mission to make everyone know this in a destructive way. Believe me, I've seen this happen many times.' Berinda explained with convictions.

I didn't know what to think about that and carried on with working with Berinda right up until the end of my first week.

Julie was married to Robert Strike and they had three beautiful children. Cassie was one year old, Bradley a few years older and Amanda a few years older than Brad. I could hear Julie's calls to Robert, who was tasked by the board to be with me during my first week.

On one occasion, I heard Julie yell over the phone to Robert, 'It's your turn to breastfeed the baby!'

Robert screamed back at her, 'How can I when I don't have breasts?'

Once my first week was up, Berinda left and returned to her TAFE job. I didn't think much about Berinda's warning until I arrived at work on my second week, and I was told that Julie wanted to see me as soon as I got in. I felt abandoned with Berinda gone and I didn't know what to expect with Julie.

Taking a deep breath, I walked into Julie's office, hoping for the best.

'Hello Julie, great to finally meet you!' I announced as I walked in and shook her hand gingerly.

'Oh yes, so you're Arthur. I didn't expect someone like you!' Julie said ambiguously.

'I hope that's a good thing?' I relied with a restraint smile.

'Looks like they finally did something right this time without me, that was strange. Very strange!' Julie said as she was looking at me up and down.

I was pleasantly surprised at how well we both got along on immediately. For someone who stands less than five feet tall, Julie has such a huge personality, it seems as if she is six feet tall.

Right from the start, I knew she was special. It wasn't only her personality—it was her knowledge and professionalism. It didn't take long before we worked up incredible respect for one another. The more I got to know her, the more she knocked my socks off with her understanding of clients' needs, her extensive networks, and her shrewdness.

Julie was working at SAS but was on secondment to NSW TAFE Consumer Support and Training Program delivering standards training to people with a disability. Although she had her own office at SAS, she was hardly ever there because she was always out doing training.

It never occurred to the board or the staff that without Julie, achieving certification in four years was impossible. From the moment I met her, I wanted her on board. She was the key and without her, I knew I was not likely to succeed. My first board meeting was in fewer than two weeks and all I did during that time was prepare my case to the board regarding bringing Julie back to SAS.

30

PNEUMONIA, 1997

Despite having a skeleton staff and a shoestring budget, I didn't panic because I knew, one way or another, I would be fine. Well, I would be fine once I got Julie on-board as my second in charge.

My efforts working with the part-time admin person and the disability trainer were going nowhere fast and they both informed me that they were calling the shots. They made it clear that, as the manager, I was to follow their lead or suffer the consequences. They didn't need to spell it out to me; their actions made it clear.

It felt like a lifetime had elapsed when finally, my first board meeting arrived. I put forward my proposal for Julie to come back and work as my deputy, and the board approved it. A huge load was lifted from my shoulders. Now I knew I was on my way to putting my plans into action.

But her TAFE boss knew they couldn't find a replacement of Julie's calibre and made it hard for both Julie and me. They stretched it out for weeks before releasing her. However, our patience paid off eventually.

As expected, she was marvellous right from the start and from that day, we didn't look back. It had been worth going through the difficulties to have Julie working with us at SAS again.

'Why is she working in that office, she shouldn't be working here!' A client called out in front of all the other Monday regulars.

'Julie doesn't work at TAFE anymore; she works with us!' I replied with a grin.

'Oh boy! That means we are going to have some fun now!' She said then turned and repeated it a few times around the office getting everyone even more excited.

It took no time at all for Julie to agree to my plans, after first making some changes to prioritise some areas of concern. She thrived with her new responsibilities. The harder we worked together developing our draft policies and procedures, the more we ignored the two part-timer distractions. However, we couldn't overlook them forever and it was the beginning of the end, unfortunately, for both of them.

We replaced them with two new staff members: a new office administration manager and a self-advocacy officer. The latter was a wonderful guy called DJ, short for Domingo Jose Palazon. The four of us made a perfect team and I was determined that nothing was going to stop us achieving our goals.

'I didn't think your new team was a good match Arthur, but with all the changed the clients' feedback haven't look so good.' The President said in front of the entire board.

'Thank you Robert, also we will soon have payroll and banking all computerised thanks to DJ. We are putting better systems in place as well as doing our best with our clients heading into the next millennium.' I explained while at the same time pointing at my team.

A few days later, I found myself in hospital with pneumonia. I couldn't believe my luck. Just when we had put together the perfect team to take SAS to its pinnacle, I became ill.

In fact, I was very seriously ill. It was still winter, and I felt terrible. Overwhelmed with weakness and discomfort, I was sick and tired of worrying about my mortality. I was ready for death and was at peace with it. I had kissed my kids, my wife, and my parents goodbye and was looking forward to the inevitable with open arms. At this point, I was fading in and out of consciousness.

'Come on honey, tell me what's up.' Helen asked with a strong look of determination in her eyes as she leaned in closer to me over the bed.

'I've had enough honey, I'm too tired and weak. I'm ready to go, it's okay.' I whispered back with a soft voice trying desperately to avoid eye contact. I explained, lying in the hospital bed sweating profusely, and staring at my multiple IV drips.

'I know honey, I know. You just rest and don't think about anything. This will past and you will be back home in no time okay. You just need to rest. I will not leave to take the kids home until you rest please. Stop with that nonsense, the kids are listening.' Helen simplified things for me before I drifted off to sleep.

One morning I woke surprised to find Julie standing there with flowers in her hand and I thought, this is my final day. I must have had four or five IV tubes all running to a vein in my neck by then because the veins in my arms were all unusable.

'Julie, is that you?' I shout while rubbing my eyes.

'Yes, I found you finally. You know how hard it was finding you. This place is huge!' Julie said out loud as she was still offloading all her bags she was currying.

'Why are you here?' I said expecting to be ignored and in total shock.

'I'm here because I care about you dummy!' Julie explained still unloading her.

'You shouldn't have gone to all this effort Julie.' I said first annoyed and then relieved almost after it came out.

'Well, I'm here, how are you?' Julie said directly into my eyes with her full attention.

Julie wasn't fazed by any of it and was just glad to see me. After I thanked her for visiting, I didn't know what to do and just lay there in disbelief. I couldn't get over the fact she had made her way over to the RPA just to see me.

That day was my turning point. Julie stayed for hours, and I found her company soothing. After she left, strangely enough, I started to feel better. Although I don't remember much about her visit, I do remember her wonderful spirit that lifted me from despair and from the horrible feeling of giving up.

As I lay there, I started to rethink my decision to avoid telling anyone from my work about having thal. Julie's reaction and acceptance after seeing me half-dead with multiple IV tubes protruding from my body gave me the courage to at least entertain the thought of telling my co-workers.

A long time ago, I had made a promise to myself to never tell anyone from my work that I had that. I had made this promise to myself just after my parents sold the family takeaway business and I had been interviewed for and offered a trainee assistant manager position in a financial institute. I had an aptitude for numbers and was so proud of myself for getting this job, as that was the direction I wanted to take. It wasn't working in a bank, but it was the next best thing, and I was on my way. All I needed to do was to meet with the board of directors before starting work the following day.

The day before the interview, I had approached my doctor at the children's hospital and asked him if I should open up about these blood days if it came up in my interview. His reply was predictable, suggesting I should always be honest with my employers, and I subsequently felt a little embarrassed about asking him.

'You don't understand doc, but people aren't like you!' I yelled back in frustration.

'Listen to me Arthur, honesty is the best policy!' Dr Bau said with a calm and positive manner.

During my interview, I was lucky enough to evade any sort of question about my health. However, at the meet and greet with the board of directors after my successful interview, the chairman of the board looked directly at my interviewer as he asked me, 'Why do you need one day off a month?'

Naïvely, I opened up about my blood days and explained my situation in detail. The next thing I knew I was being hastily ushered out of the board room. One minute I was being congratulated and welcomed by all the directors and the next minute, I was left standing alone in a back corridor, as if I had the plague.

Those few minutes as I waited alone in that dark back corridor, which led to the back carpark, felt like forever. Finally, the door to the board room opened abruptly and my new boss leaned out into the corridor, without actually stepping out of the room, and yelled as loud as he could, 'You didn't get the job and it's best if you leave now!' He immediately disappeared and the door slammed shut behind him.

Speechless, I stumbled out into the carpark. How could I be so stupid? I kept on asking myself that over and over again. I had got the job and minutes later, I had lost it. I couldn't believe it. I had wanted that job so much and, in my excitement, I had told them too much. I'd told them everything.

It was the early eighties and the EEO laws were only just getting underway and far too premature for me to be aware of them. My foolishness turned into deep despair, and I could hardly walk to my car. The board members had made me feel like a leper and they couldn't wait to eject me from the building. To combat feeling lightheaded and sick in the stomach, I started to take deep breaths.

I don't know what bothered me the most, my stupidity or their ignorance. I quickly took hold of my emotions and promised myself that this was not going to happen twice. *No way am I going to forget this.* But it was the early eighties and the less said the better. That was the best way to navigate through those difficult times.

As I lay there in RPA, contemplating days past, my thoughts turned to Julie. *If I make it through this, I will tell Julie everything.*

Before I knew it, I was back at home enjoying some peaceful days alone with the kids before returning to work. Helen was at work, having gone back two years earlier.

After a six-year break from the workforce, Helen believed that Jimmy and Pamela were old enough to handle her going back to work. My resistance was futile because we needed the extra money. My job at SAS allowed me to take the kids to school every morning. Then Helen's parents, who lived in the same street as us, took care of them after school. Helen's dad enjoyed picking them up from school and they loved spending time with their grandparents. The routine worked well for everyone, and the money Helen brought in took a huge load off both of us.

Helen had initially returned to work in the health insurance sector and was now in final negotiations for a full-time position at one of Australia's leading Bank. After almost completing her TAFE private investigator qualifications, she was planning to enrol in a degree in fraud investigations.

While in the RPA recovering from my brush with death, my doctor gave me my latest ferritin results which showed a decrease of over twenty thousand. In seven years, I had managed to decrease my ferritin level from thirty thousand to less than ten thousand. The doctors explained that they had seen readings as low as five thousand. I couldn't believe it.

'Don't get used to it Arthur, how long will that last?' Peter yelled out in front of everyone for all to hear.

'If you're like that Peter, I won't tell you my results again!' I yelled back in frustration.

'Why you screaming at me, it not my fault. I'm not the one who doesn't use it!' Peter screamed out in a condescending voice.

'Well I am using it, aren't I dickhead!' I replied in total anger while staring him down.

'I'm only saying this because I love you Arthur!' Peter screamed back chuckling hysterically.

'I give up talking to you!' I yelled back sick and tired of him constantly needling me.

My daytime job at SAS and the routine of it all and consistently using my deferral pump during weeknights had done wonders for me. My thal mates were well ahead of me with ferritin levels now hovering under three thousand, which gave me hope that I could bring mine down further too.

A few days later, I got back to work and, with the draft policies and procedures almost finished, we refocused our attention to working with our clients and pulled back from attending all our government meetings. Both the board members and staff were endlessly attending these meetings, which I thought was senseless. If we were getting ready for government certification to get recurrent funding, then we needed to adhere to our funding agreement and we weren't being funded to attend government meetings.

Almost immediately, we took our attention from systemic and government-focused issues to rethinking how we could best focus our attention on people with an intellectual disability, our clients. We had all been too busy jumping through the bureaucratic hoops that we had had no time to support our clients. It is no wonder we had no policies and procedures in place.

It was a no-brainer and this decision put SAS back on the map. We were so successful so soon, we quickly realised we didn't have the staff to cope with the huge influx of clients and we needed to pull back our networking and efforts to promote our services.

Our annual funding was enough for rent, wages, and utility costs. It covered three part-time workers and one full-time staff member to provide support for only fifty clients annually and there was no room for anything else. As we hadn't yet secured recurrent funding, we knew that any attempts to gain an increase in funding would be futile.

No matter how ridiculous I deemed the reality that SAS supported only fifty clients annually, there was nothing I could do about it. It was traumatic for me to think about the many people with an intellectual disability who could benefit from our service, but I had to put that thought aside to move forward. We needed to focus on supporting our existing clients and obtaining recurrent funding for the organisation before attempting to expand our services.

31

DR JOY HO, 2001

By 2001, my thal mates and I had settled into our newly created Thalassemia and Haemophilia Clinic on the second floor of the Page Chest Pavilion at the RPA. The Page building was later demolished in 2009 and replaced with the magnificent Chris O'Brien Lifehouse building in 2013.

By 2001, we had lost contact with the head of the department, Dr Harry Kronenberg, as well as Dr Ron Trent and Elizabeth Swinnerton, who had all looked after us since we'd first transferred over to the RPA. With an array of many weird and not-so-pleasant unused wards and empty rooms at the hospital, we didn't care where they put us as long as we had Fiona Rennison.

Up to this point, I had managed to avoid getting

a portacath inserted in my chest, I still had the problem of bad veins, and the tolerance of the clinic staff was running out. It was Fiona's professionalism that saved me, just when tempers were peaking again. Resistance was escalating and, at one stage, I demanded to see a specialist every month even for my cross matches. They all resisted putting in my IV, blaming my poor veins instead of their incompetence.

What made things worse was that I was effectively questioning people's skills by suggesting that my veins couldn't be that bad if a head nurse like Fiona could find one on her first attempt every time. She never missed and I never felt any discomfort or pain whatsoever. I was always relieved to find her there; it took a huge load off my mind to know that she was looking after me.

Occasionally, thals from other hospitals came into the clinic. They were always at least ten younger and would be there for treatment for other conditions. It was refreshing to see new faces and get to know other thals. All of them had wonderful personalities and were filled with positivity, which was uplifting. We kept them entertained with our ridiculous behaviour and impudent jokes.

'Why are you all crazy, it's never boring here with you lot.' Their comments would always be the same.

'Yes, I can't help it. I'm the life of the place. Peter said loving an audience. 'Come back any time, I'm always here!'

After discovering some of them had portacaths inserted into their delicate young bodies, I was in awe of them. They were all trying to do whatever they could to avoid making the same mistakes I had made when I was their age, even though they started on the treatment from a much young age.

We were sick of being kept in the dark and not always having access to our monthly test results, so were pleasantly surprised to be informed of a new doctor, Dr Joy Ho, who would see us all routinely. This was welcome news as we were sick of seeing different doctors every month.

Initially, when we met Dr Ho, she was just another doctor, but we liked the idea of having someone review our results on every visit. At times, we were desperate for someone to tell us what we needed to hear, good or bad, and she did that.

'How long will you be around?' Peter asked brashly.

'I'm planning for a while actually?' Dr Ho replied with a smile.

'No offence, but that's what they all say.' Peter answered back with a huge grin.

'It was a pleasure meeting you all and I'm looking forward to seeing you next month.' Dr Ho was convincing looking around at all of us before heading off to her other patients.

'We'll see doc!' Peter had to have the last word.

However, on our second visit, she took us all by surprise. We all knew the new routine regimental checks were not a coincidence and we were entering a new era. It almost felt like we were back at the children's hospital, and we were imbued with a much-needed sense of hope and confidence. Feeling much better about being informed, in detail, about our monthly blood results and having our ferritin dosage also routinely adjusted for optimum efficiency, I found myself enjoying our family outings more and feeling more confident.

For a change, we felt we knew where we were headed, with the course of our treatment plotted out for us. We knew when we got off track too. Once we had been blind and now we could see, and our focus suddenly turned to pleasing Dr Ho because we had been lost and she had found us. She had turned on the light so we could follow her out of our darkness.

———

By then, I had lost all contact with Frank. However, Ron had returned from England and was living nearby after recently getting engaged. I had lost count of how many marriages he'd had but I hadn't seen him so happy for a long time and I was thrilled for him.

After convincing my son Jimmy to join the Australia Air League a few years earlier, Ron pushed for Jimmy and me to make the move over to The Australian Air Force Cadets (AAFC). The AAFC paraded on Monday evenings at Richmond Air Force Base. Jimmy had just turned fourteen, and although I was content to have him stay at the AAL, which I loved, I left it up to him to decide. By that time, they were located at Whalen Reserve in Mount Druitt and parading at a building specifically built for them. It was a vastly different experience from congregating at the local school as Ron and I had done at Jimmy's age.

After arriving home after nearly two decades living away from Australia, Ron was back at the AAL. He had finished school in England and had joined the Royal Air Force and, early in his career, he had been tasked with helicopter maintenance during the Falklands War. He was rising in the AAL ranks and it wasn't hard for him to convince Jimmy to join. I was in awe seeing Jimmy in a uniform I loved and cherished.

But now Ron wanted us to move to the AAFC and he got his way again. He convinced both Jimmy and me to join the 336 Squadron (SQN) AAFC at

Richmond Air Force Base that year. Jimmy joined as a cadet, and I was an adult uniformed instructor with the rank of Pilot Officer (AAFC). Although the AAL will always remain dear to my heart, this was the next level and Jimmy didn't hesitate to join. It wasn't only that 336 SQN was nestled in a fully operational Australian air force base, being among a group of incredibly gifted and talented thirteen to eighteen-year-old cadets was unbelievable. After being used to parading with a dozen or fewer cadets in the AAL, we lost count at one hundred and fifteen cadets on our first visit.

The Commanding Officer (CO) who was at the time ranked Flying Officer (AAFC) was an incredible giant of a man by the name of Dennis Lockwood. He was admired and revered by and won the hearts and minds of all cadets, instructors, staff, volunteers and especially the parents.

'I hear you both are from the AAL, welcome. Please feel free to look around and I would be happy to answer any questions you have later!' The CO warmly said before heading off to start the night's first parade.

He didn't hesitate to take on both Jimmy and me. He and his marvellous team and especially the parents' volunteer group, made us feel welcome even though we were both from the AAL. The CO valued everyone's support, he had cadets and instructors from all sorts of backgrounds from the AAFC and AAL to the boy scouts and girl guides. The girls made up more than a quarter of the squadron back then and their numbers grew fast.

It was well-organised and structured, with an absurd amount of discipline and dedicated eighteen to twenty-year-old cadet instructors leading the new and aspiring cadets to superior levels. Marching was only one of a huge range of activities they offered between open and closed parades. Classes included aviation, bushcraft, adventure training, teamwork, effective communication skills and many more introductory classes. As the cadets progressed, moved up to completing a Duke of Edinburgh's award or even getting a licence to fly gliders or civilian aviation.

We were in the midst of all this with a backdrop of C-130J Hercules and Caribou, later replaced by the C-27J Spartan, aircraft routinely coming and going and doing touch-and-go landings on the runway day and night. Occasionally, both Jimmy and I were stunned to see an array of jet fighters but that didn't faze anyone else because they were used to it.

Access to the base each Monday evening became much more difficult after September 11. Base security was heightened, and all visitors were methodically checked and verified. What a time to be there! But 336 SQN managed to parade every week all the while under close scrutiny and constant surveillance.

One evening in July the following year, I had a poignant and unforgettable experience. While completing my AAFC officer training, I was staying on base in the officers' accommodation, and it happened to be the same time that the senior crew of the Nottingham were also staying on the base. On 7 July 2002, the British Royal Navy Type 42 destroyer *HMS Nottingham* had run aground near Lord Howe Island, 600 kilometres off the New South Wales coast and 780 kilometres from Sydney. The destroyer had nearly sunk after colliding with Wolf Rock due to a navigational error in bad weather following the transfer of a sailor with a medical condition. The incident resulted in worldwide media attention and much embarrassment to the British Navy, followed by a complex rescue operation and a massive repair bill that was reported at the time.

'Hey guys, the officers from the Nottingham are all here. They're all pretty much kept to themselves and away from the rest of us while making their way to the officer's mess hall earlier on. Their captain is with the base chaplain and are making their way here soon.' A voice behind me softly called out for only a few to hear.

'I'll tell you, the Nottingham captain would have been proud of the way their officers conducted themselves this evening. I guess, difficult circum stances and the huge amount of pressure they must have all been under. I'll never forget what I saw. It was an example of how exceptional professionals conduct themselves under extraordinary and challenging conditions.' The voice continued as we all got up from our bar chairs and tried desperately to get a glimpse of the officers.

'What did you see?' I whispered not taking my eyes off the officers from the other side of the huge officer's bar room.

'It was, what I didn't see, more like it.' He explained with a sincere look.

'Okay, what didn't you see.' I eagerly replied without looking back to see who was talking.

'I didn't see them drinking at all, talking out loud or even joking and messing around. In fact, they all helped each other getting one another's cutlery and food in a polite and proper manner. Then later making their way back to the officer's bar to order tea and coffees. I'll tell you all, these guys are professional through and through. My heart glows knowing these officers are our allies, guys.'

With my mouths remained opened, we were left with a lasting impression on all ten of us because although they were all professional Royal

Naval officers, they looked like normal people just like all of us who were watching them and witnessing their extraordinary professionalism. None of us said anything to each other after that, we just very felt proud.

Months later on one warm parade night, we were informed that adult volunteers were required for the base's routine mock rescue evacuation exercises on the following Saturday. After volunteering for the drill and getting the detailed brief, I didn't think much more about it until the day of the exercise when I found out how seriously the base took these drills. The volunteers were instructed to report to a designated section of the runway and told to put on our nominated medical injury and lie down in a position appropriate to the assigned injuries.

'Thank you all, now once you are all in place, I want it to look like a war zone. The C-130J Hercules will be here soon. It's fitted out with military-style rail bed bunks stacked three high that will be landing close by. Then, in a precise and proficient manner, their crew will be loaded all of us in as quickly as possible. I need you all to marvellously playing your roles in line with your nominated injuries, onto the military plane please. Okay, get in place, here they came!' The volunteer coordinator instructions were clear.

Under the enormous loud noise of the engines of the Hercules, we were all quickly loaded and locked into our bunks by the crew and the plane took off immediately.

Once in the air, a round of applause from the crew rang throughout the fuselage of the Hercules. Having achieved top marks for their efforts, they thanked us volunteers who were still strapped tightly to our bunks.

On this bright and clear day, we flew south towards the Victorian border and finally turned back at Albury. On the way back to Sydney, they gave each volunteer a turn to visit the cockpit. It was a wonderful thing to do to show their appreciation for accomplishing top marks in their exercise.

The following summer, Jimmy and I signed up for a two-week gliding training course at 328 SQN AAFC Bathurst. I was so excited thinking that Jimmy would most certainly end up enrolling into the Royal Australian Air Force after year twelve once he had flown in a glider and got a taste for flying. But my hopes and dreams quickly dissolved on the first day of flying.

I went up first and, after getting up to 5,000 feet, my tandem instructor graciously allowed me to take control and I manoeuvred the glider down to almost ground level before he took back control to land. Jimmy, however, only got to 3,000 feet before he turned green thanks to his instructor playacting several disaster mishaps before cutting the flight short.

When he got back on the ground, I didn't wait to get permission before taking him to the Bathurst Hospital. After spending over six hours in emergency for observation, they allowed him to be discharged only if I agreed to take him home immediately. I needed to keep him away from others as he had been diagnosed with a possible virus with an elevated temperature. Jimmy didn't argue and was happy to be taken home

I was shattered when Jimmy told me his flying days were over. My hopes and dreams of him joining the Royal Australian Air Force crumbled into dust. I didn't care what profession he took up as long as he was in the RAAF. Maybe moving to AAFC and exposing him to all of this hadn't been such a good idea.

By then, Jimmy was a senior cadet and was I teaching an assortment of basic classes. I was so immersed in my role, helping out any way I could, I didn't notice how well he was progressing. Then one evening, I was called over to see what was happening in one particular classroom and I saw Jimmy in control of a class full of first-year cadets.

'Check that out Arthur, there must be about forty to fifty new cadets and Jimmy has them all in the palm of his hands.' The executive officer said standing next to me at the doorway of the training room.

With an assortment of ice breakers and pop quiz activities, the atmosphere was electric with clapping and cheers that filled the entire squadron.

After taking another look around, I noticed that Jimmy was the senior cadet in the room. Then it dawned on me that Jimmy was their instructor. At that moment, all my worries about how he would cope in the real world lifted. It was as if the earth's tectonic plates had shifted beneath me, and I was now in a world I was not familiar with.

We both left the AAFC the following year, as Jimmy needed to focus on his high school certificate (HSC), and I had enrolled in university. It had been a wonderful experience for him. However, for me, the highpoint of the time we spent at 336 SQN AAFC was seeing him instructing that base aviation class, like a maestro conducting an orchestra, and witnessing how he conducted himself with such authority, charisma, and total control while having so much fun.

32

GOING TO UNIVERSITY AT FORTY-TWO, 2003

In 2003, I finally made up my mind to enrol at university and do a degree in adult education. For my entire adult life, I had wanted to be a TAFE college teacher. I had no aspirations to teach kids but teaching at TAFE was a dream of mine.

After Julie and I had completed the Certificate IV in Assessment and Workplace Training the previous year, I put some thought into achieving this lifelong goal. The problem of getting a tertiary qualification without any underpinning qualifications was bad enough, but would I live long enough to use it?

Soon after Julie and I graduated from TAFE, we were shocked to discover that she was the first person in Australia with an intellectual disability to successfully attain national qualifications at this level. This was amazing and it spurred me on to think more seriously about going for my goals. I already had four nationally recognised diplomas from McDonald's as well as my new Certificate IV qualification from TAFE. All I needed was to come to terms with that old chestnut of my mortality.

By then, both Jimmy and Pamela were in their mid-teens, healthy and doing well at high school. Helen had been a full-time fraud investigator at the bank for over ten years. It didn't matter if she worked in the city or at Parramatta, she would always be home before six p.m. to have dinner ready so we could all eat as a family. Helen had been remarkable, never faltering with this decision no matter what. In addition, she would always clean up after dinner and chip away at the housework, freeing the kids to do their homework and me to mix up my desferal drugs and use my pump.

The kids were used to me doing my drugs after dinner. However, just after Jimmy started school, I was summoned to the school one day to explain myself when Jimmy announced in show and tell in his year one class that his dad was doing drugs every night after dinner. From then on, I stopped referring to it as 'doing drugs'. As the kids got older, I waited until I'd taken Pamela to violin lessons and Jimmy to soccer training to start my desferal treatments. I particularly loved taking Pamela to violin lessons and would sometimes spend more time talking to her teacher than she had with him doing her lessons.

I don't know how it happened, but I ended up being Jimmy's soccer coach. The local soccer association made me do a course to be accredited and, after a lengthy period of getting qualified, they gave me his entire team to coach, and I had never played sports in my life. These poor kids looked up to me, all in need of direction, after the previous coach resigned because of the parents' complaints that the team was always in last place. It was early days, and the angry mob was quiet for now, but I needed saving quickly before they caught on that I had no idea what I was doing.

The soccer association didn't want to listen to me when I said I had no skills whatsoever. For some unknown reason, they thought I was being modest and the more I explained myself, the more they didn't listen. I passed the qualifications to be a coach and that's all they cared about. I only did the damn course because they paid for it, and I hadn't expected to pass.

'Angelo, guess what, I passed my coach accreditation! Bro, what I'm I going to do now?' I screamed out of the driver's car window as I pulled up at his place in total panic.

'That's great bro, well done!' Angelo replied impressed with my efforts but forgetting who he was talking to.

'Will you stop that; you're talking to me don't forget!' That's when the penny dropped for Angelo and reality hit him hard.

'Oh, shit. What are you going to do now? You're not still going ahead with it?' Angelo smile turned to horror.

'No, not me! You are bro. I need you to be the coach. I'll be the manager and you coach the kids!' I desperately tried to explain with hopeful eyes.

After spending the afternoon begging him to help me, Angelo reluctantly agreed to trial it to see how things would go and only if the kids agreed to it. As a soccer enthusiast and a brilliant strategist with statistics, especially with soccer players' strengths and weakness, I immediately I knew I had no choice but to get him to agree. We both knew it could only end badly if the mob ever discovered how little their new coach knew about soccer.

It didn't take long for Angelo to discover the real strength in Jimmy's team once we started, identifying the one girl player we had as the star player. She was the fastest and he coached them all that she would be the one to manoeuvre the ball in and out the entire length of the grounds, and if she couldn't get a clear shot she would pass it to either wing players for a shot.

After a few weeks of practising, the team was ready and that's when the fun started. Angelo was magnificent as assistant coach. The mob settled, the players learnt new skills and attendance was up. We didn't win that year, but we didn't get the wooden spoon either.

Jimmy's attention soon turned to cars, in particular BMWs, and my coaching days were finally over, thank goodness. I was grateful to Angelo for saving me and also for helping me to spend two priceless years as my son's soccer coach, even though I had no idea what I was doing.

By then, Pamela's violin lessons had also ended after I took her and her violin tutor to the Sydney Conservatorium of Music in Macquarie Street so she could sit her end of year exam. For a straight-A student, this was a walk in the park for her and I enjoyed every minute of it. I was thrilled I was around to experience such treasures. Working regular nine to five hours at SAS had allowed me to significantly decrease my ferritin levels and experiences like these were a great motivator for me.

I decided to stop carrying my heavy baggage. Since I was twelve years old, I had carried the weight of my mortality every day of my life and it was time for me to put aside that worry.

'Helen, I've decided to go to university.' I said with much conviction and determination one evening after getting home from work.

'I can see you are serious.' Helen responded with a smile before going back to getting changed out of her work cloths.

'Yes I am, what do you think?' I waited for her answer holding my breath.

'Big commitment, but if that's what you what, I think go for it.' Helen casually replied now focusing of what to make for dinner.

This decision was liberating, and I was pleased with myself for finally coming to terms with my reality. I realised I had achieved far more than I could ever have imagined and even if I died that day, I would have lived twice as long as I had expected to.

With encouragement from Helen, I finally put in my application for an undergraduate adult education degree. At the age of forty-two, I was going to university.

Once my application at the University of Western Sydney was successful, I discovered a wonderful thing called recognition of prior learning (RPL). Mature age students are offered a maximum of sixteen units RPL off a degree comprised of twenty-four units. Sixteen units are equivalent to three diplomas that are nationally recognised.

With all my qualifications and my current senior management position, I was eligible to claim the maximum of sixteen units off, which meant I only needed to do eight units to complete my bachelor of adult education degree. That year I did four units, studying after work and at night, and completed the remaining four units the following year.

My first year was exhilarating but it happened to coincide with my treatment for hepatitis C. Back then, all thals of a certain age became infected with hepatitis C because there had been no screening for hepatitis C in the donated blood we had received during our adolescent years. We were informed that if treatment wasn't immediately undertaken, we would most certainly develop liver cancer within the next ten years.

Peter, George, and I all commenced our twelve-month treatment despite the side-effects of memory loss, migraines, intensive sweats, stomach pains and many other unpleasant things. This didn't make life easy, but it didn't put me off working hard that first year balancing my full-time job, evening classes, and studying on the weekend.

'Arthur, as you are aware by now. Unfortunately, Peter and George only got into their second month before things got so bad for them that they had to terminate their treatment. Dr Ho and I are extremely disappointed with this result. Now regarding you, we have had no indications to terminate your treatment. Your results have been very position so, tell me, how are you feeling?' The new registrar for Dr Ho explained as delicate as she could.

'I'm so upset that the guy's treatment was terminated but honestly, I'm feeling fine.' I replied now feeling I missed another bullet.

I managed to finish my entire treatment that year and I was cleared of hepatitis C for the first time since commencing regular blood transfusions. I also successfully completed my four units at university that year, surprisingly averaging distinctions.

The boys ended up trying another type of treatment which was also unsuccessful and were both advised to wait for a new treatment.

———

'Helen, what happened, are you all right sweetheart?' I called out from the bedroom door after noticing Helen in a ball on the floor next to the bed.

'It's dad, things are getting worse. He doesn't have much time left.' I felt Helen's cries deep inside that broke my heart for her but all I could do was hold her closely.

Then with the death of my father-in-law John Didaskalou, the following year in May 2004, things were even more difficult as I tried to give back to

Helen some of that wonderful support she had provided me since we'd first met.

After her father was diagnosed with lung cancer. In the nearly two years leading up to his death, I was able to support Helen's parents by taking them to the hospital every second morning for her dad's radiation treatment. I had hours accumulated at work, so I was able to start late those days after ensuring her parents were both safely back at home. Then, at her dad's funeral, again I didn't disappoint Helen by not leaving her sight and doing my best to comfort her in every way possible. I was so grateful to still be around to support her when she needed me the most.

Since moving to Minchinbury in 1990, my in-laws had been our neighbours and we loved having them so close. They were both special people, but Helen's dad had that special something that meant everyone wanted to spend time with him. He was a wonderful storyteller and always had time for everyone. Helen's mum, always referring her as mum, always kept busy in the kitchen or around the house never taking a minute to rest. Her dad's barbecues were legendary and when he had them on special occasions you were lucky to get a seat at the table and you wouldn't want to be late.

Helen's parents never ventured far in the car but when they did, it would usually only be to the shops, to my parents' place or their other relatives nearby.

'You know son; I want you to continue my barbeques if I get too old or something happens to me okay. I know you have been keeping a close eye on me when I do them, I've noticed. Okay, I need you to do that for me please.' Dad said out of the blue one weekend before focusing back watching the footy on the TV and leaving me speechless that he noticed I was observing him for years.

Those days when Helen's dad and I shared a coffee and talked about philosophy left a big impression on me and I cherished those moments with him the most.

After the funeral, Helen's mum did her best to cope but that spring in her step was gone and we all missed him, especially Helen.

33

MASTER OF MANAGEMENT, 2005

In early 2005, after successfully completing my Bachelor of Adult Education degree and securing my first part-time casual teaching position at TAFE Nirimba in Quakers Hill, I started teaching the Certificate IV in Business Administration on Wednesday evenings from six p.m. to nine p.m.

I also immediately enrolled in a Master of Management postgraduate degree. Now that I had a degree with a qualification to teach adults, I wanted a postgraduate degree in management because that's what I was passionate about. I knew anything less; I wouldn't have had a hope in hell of accomplishing anything near a pass mark at this level.

I was on a roll and, based on my recent success with my undergraduate degree, after realising the master's degree only comprised of eight units, I stupidly decided to do four units the first half of the year and the remaining four units the second half to complete the entire master's degree in one year.

So while working full-time at SAS, I attended evening university classes on Monday, Tuesday, Thursday, and Friday and also taught at TAFE on Wednesday evenings. Balancing all this was incredible but what was even more incredible was that I kept on using my desferal pump although it was getting later and later in the night before I started it.

Unfortunately, I had disregarded the fact that the eight units were at a much higher level than I first anticipated. However, after reassessing my options once I received my results, which showed I barely managed to get a pass mark in all of my first four units, I decided to keep going.

To attempt to ease the pressure, I would repeat to myself 'Ps (pass) means degrees'. This was so wrong in so many ways, but I failed to understand it at the time. I was quickly burning out and I sought justification to get me through the year as I received only passes, no matter how hard I tried to improve my marks.

Things got so bad for me that I complained to my professor in one particular unit called Organisational Learning and Development after he put me in a group with four other students. I struggled to get my work done and juggle everything going on in my life and I'd had enough of helping others and carrying other group members. I demanded to work on my own and not be put in a group. The average age of the students in this class was early twenties so they were in a different place from me.

'Okay, then can I talk to you privately outside Professor?' I politely asked with a death stare.

'I have no time for that Arthur, your options are clear so make a choice and please don't disrupt my class again.' The Professor replied before going back where he left off.

After he'd finished with me in front of the entire class, he demanded I make a decision, and quickly. My options were to either leave immediately and look for another unit in which to enrol or stay and do as he asked. Too tired to look around for another unit, I decided to stay and miserably slunk into the group he selected for me.

'Hi Arthur, I think I'm in the wrong class as well mate.' The students next to me whispered without drawing any attention fearing the Professor's response.

In the past when working in groups, I had found that I was always left doing the hard stuff. By then, I was sick of it and wanted to do things on my own.

After entering the group kicking and screaming,

I ended up cherishing every moment of that class. I learnt to not underestimate the power of working in a group. I also gained an understanding of how to implement working groups and discovered that if one is set up correctly from the start it can be powerful and effective.

'Hey guys, since we all don't know each other's strengths and weakness, what about we do that before we start with our working groups?' I suggested in trying to kick off my groups' discussions of methodology.

It wasn't long before the discussions turned to deciding on who was doing what. The dialog was breath taking and the ideas were amazing. It was incredible how everyone equally took on a task effortlessly and with much enthusiasm. Suddenly, I knew from this early stage of just meeting the group, I was in amongst a special collection of people the likes I have never seen before.

Our group developed a DVD for our final presentation, for which we achieved a high distinction. The unit was broken up into three parts and that high distinction contributed to giving me a 66 Credit average for the unit. This experience helped me to be more open-minded with my approach to everything.

To my surprise, in the final days of one of my other units, my professor invited me to be a casual academic and start teaching undergraduates the following year. At first, I thought she was joking, but after she explained it would be no different from my TAFE teaching, I didn't hesitate and immediately accepted.

I had undertaken my master's degree in management to be more proficient as a manager and had never dreamed of teaching at university. But my hard work and perseverance paid off in so many ways and I was pleased I finished what, at the start, had seemed to be impossible.

By the end of the year I had completed my master's degree and the following year I was teaching both at TAFE and University. I started off teaching only one tutorial unit at the university with up to thirty students, but it was great.

At the beginning of the unit, I started my classes by announcing the three secrets of how to get a pass mark. I hoped they would be interested in distinctions, but these were first-year undergraduates, and you could hear a pin drop every time I said this. I had their attention and from then on, they were mine.

'Attention ladies and gentlemen, all you need is to: 1) do the minimum reading, 2) attend classes and 3) take part in all online and in-class activities!' I explained as loud for all to hear while noticing they looked somewhat disappointed by the time I finished.

However, in hindsight, they usually realised that although it wasn't much of a secret, sometimes the obvious needed to be pointed out.

My TAFE teaching the previous year gave me the confidence I needed those first few days, and I was soon into a new routine with my full-time SAS work during weekdays, TAFE teaching on Wednesday evenings and teaching at university on Thursday evenings. For a while, it felt as if I was still doing my master's, but I was getting paid for it.

I enjoyed that first year of teaching at university, even though the year had not started well for me from a phone call from RPA, in January, while holidaying in a beautiful cliffside apartment at Manly.

'Hello Arthur, I'm one of the doctors in the cardiology unit of RPA, I need you to come to the hospital as soon as you could please, regarding the results from your last check-up. Would tomorrow be good for you?' The unfamiliar voice asked politely.

'Tomorrow is fine, let's make it after three p.m. would be good!' I replied not concerned.

'In the morning would be better, if you can make it please?' The voice said more firmly.

'Yes sure, what about 10am?' I answered more

concern now that it left me puzzled after the phone call.

'Helen, we need to cut short our holiday, and dropped off Jimmy and Pamela at home tomorrow morning and go straight to the RPA before 10am.

'Why, who was that and what's up?' Helen asked with more white in her eyes showing than ever before.

'That was a doctor from the cardiology unit at RPA wanting to talk to me about my result from my last heart scan.' I replied trying hard in lowering my voice and not to be too concern.

'Are you all right honey?' Helen asked now holding my hand.

'I don't know; well see tomorrow I guess!' I jokingly said with a chuckle trying very hard to not show I was worried.

'Arthur Bozikas, the surgeon is ready to see you now. This way please.' The reception called out before escorted both Helen and I into the surgeon's office.

'Hi doc, why am I seeing a surgeon?' I asked confused while staring at the surgeon's huge display of qualification on the wall in their ornamental frames.

'I'm sorry for the late change but after a closer review, the unit decided to immediate recommend surgery to repair your thoracic aortic aneurysm.'

The surgeon said as he sat behind his opulent desk and opened up my file.

'I have a what?' I ignorantly called out.

'An aneurysm is a weakened area of an artery which results in the blood vessel wall bulging like a balloon from the pressure of the blood pushing against the vessel wall. As an aneurysm grows, the blood vessel wall becomes weaker and if it ruptures or tears, it can cause life-threatening internal bleeding. Your aneurysm was located on your aorta which is the major blood vessel that feeds blood from the heart to the body. Due to the subsequent rapid blood loss, a rupture of a thoracic aortic aneurysm is likely to be fatal.' The surgeon explained to deaf ears.

'What size is my aneurysm?' I asked pretending I know what I was talking about while Helen kept holding my hand.

'The diameter of a normal healthy thoracic aorta is about 3.5 centimetres, and your aorta was currently variable in size and up to almost 4.9 centimetres and surgery is recommended once an aneurysm reaches five centimetres, as the likelihood of rupture increases. I would like to book an operating theatre for surgery to insert a metal mesh coil (stent) into your weakened aorta because aneurysms generally expand rapidly when they reach the size of yours. Unfortunately, it is also necessary to replace your heart valve.' The surgeon was cold and clear with his recommendations leaving us both now terrified.

'Sorry but things are moving too fast, and I need time to take it all in.' I blurred out without thinking what I was saying.

'Sure, I understand but we don't have much time. As I explained, the fear of a rupture increases dramatically at the size you're currently showing. If you like, I can recommend a second opinion?' The surgeon kindly suggested.

'Yes please.' I immediately replied as I turned and faced Helen with a smile.

'It would be my pleasure and I'll be happy to referral you to Professor Richmond Jeremy, a cardiologist and a specialist in this area.' The surgeon happily replied as he wrote down his details on the surgeon's business card.

We didn't need to wait long to see Prof. Jeremy, and after a very thorough examination, he ordered additional scans and other related tests. After undergoing a battery of scans and tests, we returned to discuss the results.

'It's no great surprise here, Arthur, these results matched those of the initial results.' Professor Jeremy explained assertively.

'So back to the surgeon is my next step, I guess?' I relied gloomily

'Not necessary, I would like to wait a further six months and see what results we get then if you are interested?' The Prof's suggestion was not what we were expecting but was exhilarating to hear.

'Really, come back in six months!' I needed to repeat it just to make sure I wasn't dreaming.

'Yes, come back in six months but we need to do the tests all over again.' The Prof. was clear.

'Yes, of course. That's a great idea!' My excitement was overwhelming.

However, we were happy to hear his suggestion to wait six months to see what the results showed before deciding to operate. I welcomed that idea with open arms and was pleased to have him as my cardiologist.

Six months later, new tests results showed no changes in the aneurysm and yet I was surprised that Professor Jeremy suggested waiting a further six months. I understood why he had given me the option for the first six months, but I was stunned he gave me another six. I had been prepared to go to battle and ask for it, if by miracle, I presented with the same results, but then he beat me to it. He was bold and daring but always remained cool, calm and collected; he was a true professional with impeccable instincts.

This had a massive influence on me and, even though I was on the edge of my seat, confused as to why the aneurysm had not rapidly expanded as the other surgeon had expected, I had confidence in Professor Jeremy and left his office to wait for another six months. Helen and I had already been fortunate to be on the receiving end of several miracles and we were in no doubt that we were graciously being granted another one.

34

RON BERTRAM, 2006

Since Ron had returned from England, he had got married again, had two beautiful kids, and moved into a lovely house in the Hills area in Sydney's west. His parents and two brothers soon followed and settled in the Southern Highlands, south-west of Sydney.

By then, Ron was truly involved in the senior side of the AAL in the evenings and running his business as an instructor pilot at Bankstown Airport. On many occasions, he phoned me and told me he had a free morning or afternoon and invited me to go flying with him. Whenever possible, I jumped to take up his offer. Over the years, he took me from the city (Sydney) to the Blue Mountains and right through from Wollongong to the Central

Coast of New South Wales. It was amazing spending all that time with him, and we talked about flying to our heart's content.

We were in contact with each other almost every day, so when I got a call from him on the morning of April the fifth, I didn't answer it because I was in a meeting. After a busy day at work, I rushed over to the Parramatta campus of the university to teach my six p.m. class and forgot to play back the message Ron had left for me earlier that morning.

When I finally arrived home at almost ten pm, Helen began to heat up my dinner and then she asked, 'Did you hear anything about Ron's plane accident? It was all over the evening news.'

'What news, Ron's plane accident. Oh no, please god, not Ron!' I shout out in total horror.

My heart was pounding in fear, I made a dash for my mobile that I had left in silent mode at the bottom of my teaching bag. There must have been a dozen missed calls. When I played back the latest message, I was horrified to hear that not only had Ron been involved in a plane crash, but he had also been killed. I had eleven more messages waiting but my mobile phone fell from my hand and dropped to the floor.

I tried to make some sense of it. How could that have happened? When I turned and saw Helen's face, that's when I knew it was true. It didn't make sense to me. He had impeccable safety procedures and would rattle out flawlessly and routinely what to do if his plane engines failed or stalled. I had spent hours quizzing him and testing him on a huge range of safety-related issues and he never hesitated to respond with the correct procedures. In addition, for the past two years, he had been the president of the Aircraft Owners and Pilots Association (AOPA).

Ron's message on my mobile phone just didn't make any sense. This couldn't be happening.

The late-night news started on the TV and there it was in huge headlines. I burst into tears, shattered beyond belief.

After talking to my family for a while, they encouraged me to listen to all my phone messages. Most were from an assortment of different people calling to inform me about Ron's crash. But the last call was from Ron. It was the call I had missed while in my morning meeting. I was distraught after hearing the message he had left me, 'Hi, mate. It's not urgent. You know I love you!'

It wasn't until after his funeral I found out the truth of how Ron died. The funeral was packed beyond capacity with all sorts of people paying their respects, from Dick Smith to AOPA board members, a huge contingent from the AAL, government dignitaries and countless other distinguished guests. Ron's family was overwhelmed by the support but were too grief-stricken to be able to fully appreciate everyone's commitment to be there on the day.

As I gave the eulogy, I looked into their shell-shocked eyes as they struggled to hold themselves together. Afterwards, I supported them by driving them home.

Back at Ron's house, I started telling them funny stories about Ron and also some of our misadventures as kids, which helped us all out of the dark place we were in, at least for a short time. But the jokes didn't last long and before I knew it, we were all totally drunk. Thankfully, Helen stayed sober and managed to get me in the car and take us home.

In the days and weeks after Ron's death, the sun seemed to shine less brightly, but after a while, we all started to get back to the swing of things. Eventually, I stopped re-playing his phone message because it wasn't helping me. What did help somewhat was getting an understanding of what happened to cause my friend's death.

'Arthur, Ron had been testing a light plane, a Lancair 360 kit aircraft. A report into the crash released by the Australian Transport Safety Bureau (ATSB) said kit planes were 2.5 times more likely to crash than similar commercially built aircraft. This particular aircraft had suffered a wheels-up landing with its previous owner.' Ron's young brother paused as tears streaming down both our faces.

'Ron, who had been a licensed engineer in the British military as you know, had made repairs to the aircraft during the twelve to eighteen months before the crash. He was also a very experienced commercial, light, and ultra-light pilot. The accident happened about 2.40 p.m. while he was practising touch and goes, in which he almost lands the plane before taking off again.' Ron's brother then broken down and didn't want to talk about it any further on one of my last sober visits at Ron's parents place in the Southern Highlands.

Once he handed me the report before starting back up with his drinking, I couldn't help reading an Airservices Australia spokesman comments to myself, 'He did a steep right-hand turn from a height of about 380 metres before hitting the ground upside-down at a high speed.'

Learning that Ron had been flying a kit aircraft rather than a commercial light plane cleared things up for me to some degree, but he was still gone, and I was still grieving.

35

SEVERINO SCARFO, 2007

By early 2007, I was lecturing as well as teaching my regular tutorial classes at the Parramatta campus of the university. By then, I often had three hundred students in one lecture, which was not normal, and it's not for the faint-hearted to take control and hold on to the interest of that many students gathered in a lecture theatre.

When I started, lecturing wasn't a problem for me. I had the advantage. I knew the subject matter. Furthermore, I had developed confidence while teaching at TAFE and I had both my undergrad and postgrad experience under my belt. The difficulty for me was coming to terms with the huge number of people who walked out during my very first ever lecture.

I kicked it off with such confidence and had all the undergrads in the palm of my hands. Questions and answers were increasingly being hurled at me from all directions. I was on a stage, wired for sound and standing in front of two massive presentation screens like motivational speaker extraordinaire Tony Robbins.

The time allocated to my class was an hour and a half, but I could have gone on for hours. I was on fire and at first, I wasn't bothered when I noticed one person get up from the front of the lecture theatre and walk out. Even when a few more followed, I kept my control because I still had everyone else's attention. It wasn't until the walkouts continued and by three quarters through my lecture, almost half of the three hundred students had exited the lecture theatre, that I found myself feeling gutted. By the end, there were only about fifty students left from the original three hundred.

What just happened after it started off so well? The lecture theatre then quickly filled up for the following lecture, so I packed up my lecture notes and rushed out before the next lecturer arrived. After making my way to my car, I sat there for a while before driving home reflecting on what I could have possibly done to cause the mass exodus. When I drove off a few minutes later, I was still confused about what had just happened.

It wasn't until the same thing happened in my lecture the following week that I had the brilliant idea to pause and walk over to ask one student why they were leaving.

'Excuse me but can I ask why you are leaving my class?' I asked politely not looking forward to the answer.

'We are trying all the lectures to decide what to pick before the end of the month, thanks Professor.' The young students both answered at the same time in total unity but left me confused why they called me Professor.

That's when it all made sense and I realised I had nothing to worry about.

I had forgotten that students had the first four weeks of each semester to select their subjects before they were locked into their choices. The students were leaving my lecture to check out other subjects before their time to select their classes ran out.

After the first month, student numbers settled back down to a good regular turnout, never dropping under half again, and they remained steady until the end of the semester.

———

In the same year, I was informed by the staff at the Parramatta campus that the university was advertising for an executive officer for the students' association. Aware that I was working full-time as an executive officer for Self Advocacy Sydney, they encouraged me to apply for the position. I did and when I was informed I had been successful, I did something I thought I would never do—resign from Self Advocacy Sydney. We had achieved certification in 2001, got our recurrent funding in place and things were not progressing as fast as I would have preferred after a decade in the position.

Excited about my new opportunity, I didn't stop to think about what would happen to SAS or how would they cope without me and, after finding my replacement, I took off to expand my horizons.

As executive officer, I was responsible for the University of Western Sydney Students' Association (UWSSA) across all six campuses: Hawkesbury, Penrith, Blacktown, Parramatta, Bankstown,, and Campbelltown. Specifically employed by the UWSSA to consolidate all six campuses of the association, I needed to manage over $1.7 million worth of staff redundancies (twenty-seven full-timers and forty-two casuals) by wind-down and over $2.2 million of students' association overheads for the purpose of ceasing operations by end of my eighteen months contract.

I was required to execute all the staff redundancies prior to commencing liquidation measures on behalf of the university. I worked with the office of the deputy vice-chancellor, Rhonda Hawkins, and was graciously recognised as encompassing a high level of integrity and compliance to systems and standards and an aptitude in risk management, business savvy and transparent financial reporting skills.

My job was made clear by an independent financial audit carried out by UWS to investigate inappropriate and misused students' association funds. I believe my pro-active approach to relationship building with regulators in accordance with UWS directives and my ability to manage internal and external compliance audit processes greatly assisted in these volatile and delicate circumstances.

The gravy train, after years of misappropriation of UWSSA funds, had finally pulled up at the station. Voluntary Students Unionism (VSU) also came into full effect at the beginning of 2007 after changes to federal legislation. This meant that no longer were student union fees automatically deducted as part of student fees. Instead, students were given a choice whether or not to contribute and join the student union. Without this guaranteed revenue stream, the UWSSA could not continue to operate in the same manner it had previously. Operations of the old students' association at all six campuses were wrapped up and the entire operation was moved to under the umbrella of the university.

This whole process was challenging, especially for the staff. I saw first-hand how devastating it was for many of those beautiful and passionate staff members to lose their jobs. Some of them had worked for the association for decades. For some, it was like losing an arm. It was heartbreaking and my first priority was to arrange counsellors.

As for the students, well, most of them were filled with hot air with their demands and pointing the finger to deflect the damage they caused.

Occasionally, I came across some gifted students who were blessed by God with superpowers when it came to academia. One student in particular, who was the president of the UWSSA at the time, was exceptional. While working with him on a report for the university, I was speechless on one occasion at how blasé he was when I mentioned to him that his writing was extraordinary.

'Congratulation Arthur, the panel would like to offer you the position of executive officer of the students' association.' The young president said with a smile.

'Thank you, I'm thrilled for the panels' confidence.' I replied with a bigger smile on the day I was first offered the position.

'So are you still lecturing and teaching tutorial classes here at the university?' He strangely laughed and paused before finishing off. 'Tell me, which campuses?'

'All the campuses.' I answered back smartly.

A few years later when I found out the young president had dropped out university to work in a public library. I didn't understand why anyone so young with such a superior intellect would do that, but he must have had his reasons. He was a rare individual and I was lucky to have known him.

———

The inevitable finally happened on 30th November 2007. The thing I had been dreading since I was twelve years old reared its ugly head and snuffed the life out of one of our thal mates. Severino Scarfo, who had recently turned forty-nine, died as a result of iron overload. Sev had been waiting for it and we all knew of his recent health complications and that he was deteriorating, but it still shocked us all to the core when it happened.

The last time I saw him, just before his death, I took my son Jimmy with me so he could help Sev with his computer. Sev was living back at his parents' place, and I could see the distress in their eyes. They knew the end was near.

Mr and Mrs Scarfo were gorgeous-natured Italian people who lived for their only child. They never complained and were always there for him in every way possible. Over the years, they'd had many clashes with Peter, who had frightened them both by announcing that Sev was moving out to live on his own again or he was moving overseas or getting married. Peter was practically part of the family and loved the confrontations and would never give Sev's poor parents a break. George and I didn't have any part in it and Sev's parents always welcomed us into their beautiful house in Five Dock.

On this occasion, I made a mistake taking Jimmy with me to visit. He was twenty and although he didn't say anything, it was distressing for him to come face to face with a shirtless Sev on that warm midsummer day. Seeing his frail body with a large pacemaker bulging from underneath the skin on his chest was confronting.

'Sorry, but can we make it make it another time guys, I'm not feeling well today!' Sev explained in the kitchen staring into space as if both Jimmy and I were invisible.'

'Yes of course, no worries. Just call me anytime mate.' I replied with a smile but I felt awful for him.

We didn't stay very long because Sev wanted to rest and had no desire for us to work on his computer. After spending some time with his parents we soon left, and we must have prayed the entire way home for a miracle.

The next thing I knew, Peter, George, and I, and all our families, were gathered together at the Five Dock Catholic Church for Sev's funeral. Under my breath, I whispered, 'It has begun.' The boys looked at me but didn't say a word. I had been waiting for this since I was twelve and now that it had finally happened to one of my thal mates, strangely, I found myself no longer frightened.

At twelve, they'd given us until early adulthood to live. But in our early twenties, we had started on treatment using desferal and now I was forty-six and Sev had made it to forty-nine. Although I was very sad for him, I started thinking that if I were to go any time soon I would be fine with it. Ever since I had put aside all my fears and decided to go to university in my early forties, I was at peace with myself at the thought of dying.

Our children were now adults, with Jimmy studying Marketing and Pamela completing her double degree in Law and Arts. Helen was still working full-time as a professional fraud investigator at one of Australia's big Bank, a job she had a passion for and excelled in. With wonderful friends who were experts in their field like Tina, Paula, and Erica, who were real shakers and movers within the bank, Helen was in her element. Yes, I was ready to go.

From then on, things started to feel better.

36

ROBERT STRIKE, 2011

In November 2011, I received a call from Julie Strike telling me SAS's president and husband Robert Strike has had a stroke and he was in hospital recovering. Although they were now in the process of a divorce, they still remained very close because of work and the kids.

This stop further discussion about SAS, I refocused our attention on Robert.

'Arthur, you must come to the hospital, Robert is asking for you!' Julie was now getting distressed but clear in her instructions.

'I'm sorry to hear Julie, don't worry, I'm on my way now. Meet you there.' I reassured her but was speechless after Julie's call.

I was shocked and quickly raced over the same day to visit him, only to discover him sitting up in his hospital bed joking and laughing with all his visitors. The stroke had caused partial paralysis to one side of his body from head to toe but he was showing signs of a remarkably speedy recovery and regained more and more function the more physio he did.

It was a circus in his hospital room and the more I told him to sit back down in his bed, the more he sat back up joking with everyone. It was amazing and a huge relief.

After Robert eventually settled down, it was great to catch up with the entire gang from SAS who were all there visiting him. But I was shocked to find out how bad things were at the office. *No wonder Robert had a stroke.*

My replacement had resigned over a year earlier to pursue further studies and Julie had been acting executive officer since then.

I was shattered, not because of what had happened to Robert, but because of what Julie was telling me about SAS. Before I left the hospital, I promised everyone, especially Julie, that I would meet up with them all at SAS the following day.

It wasn't until I sat down with Julie at the office the next day that she really opened up about everything.

'You did come; I didn't think you come!' Julie brashly said while rolling her eyes.

'I said I would, now tell me everything.' I replied as I sat next to her in the board room at the office.

Julie did hold back and began a two hours' marathon without pausing for a break. I found out since 2007, things hadn't progressed much because, soon after I'd left, they introduced a sub-committee governance board (SCGB) into the SAS governance structure because they felt SAS would be better managed this way, taking the responsibility off the executive manager. The idea was also that SAS could maintain an executive board of management of people made up exclusively of people with an intellectual disability by having the SCGB, a board comprised of people without a disability who were working within the disability sector, who could make critical decisions before they were tabled to the SAS executive board of management.

Apart from having two separate boards that never met being absolutely ridiculous, the executive manager's hands were tied in this scenario. Ironically, SAS had become a puppet controlled by the SCGB, yet they had been funded since 1986 to provide self-advocacy programs. Julie, in her position as acting executive officer, had become a ping pong ball who had no say whatsoever about what SAS did.

When I'd first met Julie at SAS in 1996, we'd both dreamed of one day having a person with an intellectual disability at the helm at SAS as executive officer, but not like this.

The tipping point for me was when Julie told me that she'd taken a bus and train one day to meet up with the chair of the SCGB after exhausting all efforts to try to meet them at SAS. She took a day off to go there to tell the chair of the SCGB that she could not do any more training for the chairperson's organisation (Julie was constantly training transition to work disability clients for this other organisation that the chairperson of the SCGB also happened to run) because she wasn't getting any of her work done at SAS.

Her response was that if Julie did not finish training all the chairperson's organisation's clients, she would report her to the board.

I was infuriated to hear about this blatant conflict of interest and how Julie was being treated. It was beyond belief the way this person was using Julie and SAS to conduct free disability self-advocacy training for her own multimillion-dollar disability organisation.

I knew I needed to get back to SAS quick smart to give them the support they had given me when

I'd needed it upon entering a sector I knew nothing about. Disability funding has always been limited across the sector and SAS's shoestring budget has been consistent from the beginning, only ever enough to pay the rent, basic running costs and to cover wages for one full-time person and three part-time staff.

Not-for-profit managers are a special kind of breed, and they learn quickly how to stretch their limited funding to maximise their clients' experience, to ultimately enhance the quality of life for their clients. With a chairperson who was forcing Julie (and SAS) to provide free client training to a huge well-funded organisation, it's crystal-clear what kind of people were on SAS's sub-committee governance board.

Julie was exhausted and when I found out her mother was in a critical condition living in a nursing home near Nambucca Heads and Julie could not even take off time to visit her, I decided to go back to SAS.

To take on this huge task, I needed to be focused. To do it properly, I knew I would need to drop both my much-loved TAFE and university teaching. Although this gave me much agony at the time, I knew I couldn't help SAS and continue to teach.

After my contract as executive officer with the

UWSSA had ended after my eighteen months' contract, I had picked up more teaching work which was getting out of control and the amount of marking required outside of teaching hours was killing me. It was a hard decision to make because I loved teaching and the money it provided, but I decided that SAS needed me more.

When I found out that the executive board intended to make me go through the full interview process again, I nearly changed my mind. I couldn't believe that, after I'd told them I was interested in going back to my old position at SAS, they still advertised it. I was gobsmacked.

Then I remembered why I had left. Dealing with people with an intellectual disability was exhausting because they worked by the book, but at the same time, it was hugely rewarding to see them rule the world.

On the day of the final interviews, I was furious when I saw four other candidates waiting, but I plastered a smile on my face and walked in to face the panel. I was shocked to see bloody Robert Strike sitting there smiling at me with his crooked grin. I couldn't believe he was well enough to be there, but there he was.

Someone inexperienced in working with people with an intellectual disability might have felt reassured or more comfortable upon seeing him there, given our positive working relationship in the past, but I knew better. I was going to have to work hard to secure this position.

When it was my turn to be interviewed, I didn't hold back and I knew the entire panel and the support people would love it, if I told them what I think instead of answering their interview questions. I hoped maybe even Robert wouldn't be able to hide his approval.

'Thank you for this opportunity panel. After the seriousness of the current situation that SAS in now in, I think it's best for me to give you my assessment of how things are and not answer any of your interview questions at this point. You can decide if this was accessible after I finished. If you give me the job, the first thing I would do is to promote Julie to self-advocacy coordinator and give her three months' leave, effective immediately. The second thing on my agenda after the executive board employed me as CEO, and not as an executive officer, would be to disband the SCGB. Then I would retrain the new staff to realign their job descriptions with our funding agreement to ensure everyone in the organisation was travelling on the same path.' I started off without a whisper from anybody.

I didn't talk about my qualifications or disability sector experience, nor did I answer any of their questions along the way because I told them they had lost all credibility. It didn't go down well when I told them they couldn't lead without having a true account of what was happening on the frontlines.

However, I told them if both Julie and I attended the monthly meetings, the board would regularly get first-hand accounts of how things were going, instead of hearsay through reports via a subcommittee they never talked to or met with in person. This way, they could steer the organisation back from the brink of disaster, rather than trying to lead with out-of-date and possibly incomplete information.

I finished by telling them they had a current client complaining it had taken almost six months for a staff member to contact them after huge numbers of messages had been left over this period.

I didn't tell them I intended to give up my precious teaching or that I owed them in any way. To me, they were an old friend who needed help and I was determined to make them see that I had their back.

When I got the call from Julie to inform me I had got the job, I was surprised to hear she had changed her surname.

With Julie's divorce from Robert finalised, she reverted to her maiden name of Loblinzk. She was now dating Joe Refalo and Joe had taken up a seat on the executive board of SAS in the position of treasurer.

Before I left SAS in 2007, Joe was a close neighbour of Julie and Robert. Strangely, it was Robert encouraging Julie to go out with Joe after their split up. Both Julie and Robert remained very close after their split up and have remained even closer after their divorce.

The surprises didn't stop there. I was floored and happy to find out Julie had gained her driver's licence and recently bought a car of her own. No wonder she was mentally exhausted, and that is without her mentioning anything about what was happening with her children. But what she was dealing with at work especially due to the SCGB and her not having the opportunity to see her dying mother was no joke.

I knew Robert got the kids after the divorce. I never could understand that, but it worked well for all of them. The support was well in place by then and the children didn't want it any other way.

I prepared myself to start back at SAS the following month, in early January 2012. I had ended all my teaching responsibilities by then had my action plan ready to do battle.

Finally, I was back at Self Advocacy Sydney, my work was going well. I managed to soon get on a few local council disability access committees after work hours and also a four-year term with a national advocacy pcak body committcc called Disability Advocacy Network Australia (DANA).

I was so busy I didn't have time to think about missing my teaching. It was hectic yet also rewarding as I immersed myself back in the disability sector with a vengeance. I needed to get busy, not only for the sake of SAS, but I also needed to get my mind off my thal mates, whose health was fading before my very eyes.

37

GEORGE LAMPITSI, 2015

When Peter, George and I were first encouraged to start treatment for hepatitis C in late 2003, our specialist predicted that if we decided not to go ahead with the treatment, we would get liver cancer within ten years.

After I had received my treatment and was cleared of hepatitis C, I didn't think much more about it. But due to the dreadful side effects, Peter and George had been forced to abort their treatment. These days strong medication can completely clear individuals of hepatitis C in a ridiculously short period with minimal side effects.

In 2013, when both Peter and George were both diagnosed with liver cancer, we were all rudely reminded of the words of that specialist. I

couldn't believe, exactly ten years later, his prediction had come true. Although the boys took it well
and we all hoped for the best, we didn't fully understand the ramifications at that time. However, I was
relieved that they both decided to go ahead with aggressive and invasive treatment because, up until
then, the treatments had failed to clear their hepatitis C.

About two years earlier, in early 2011, the three
of us had started regularly meeting together again
for lunch between monthly blood days. We had
started joking around about how great it was that
we were now getting old-age related aches and
pains. We rotated locations so if it was Peter's
month, we would go to either the Eastern Suburbs
Club or Watson's Bay for seafood. If it was George's
month we would go to the Bankstown RSL Club
and for my month, the Blacktown Workers' Club.

I don't know how we re-started our regular
lunches but once it happened, we hardly missed
one. It was never dull or boring and after every
lunch, we couldn't wait for the next one.

Re-establishing our routine also reminded us of
when we used to meet up in our single days which
was strange because, back then, none of us had expected to live this long.

Since the boys' diagnosis of liver cancer, I didn't
expect our lunch meetings to continue, but that

didn't faze them, not one bit, and we continued to meet up, to my delight.

After George and Debbie divorced, George had practically raised his son Christopher into the wonderful young man he is on his own. Raising his son and working as a Telstra technician were big parts of his life, but George also had a knack for repairing computers and both Peter and I loved him for it. Peter loved being a Mr Mum, and even though his two beautiful children, Ally and Michael, were now adults, he did not need to work because his brilliant wife Amanda worked full-time within the insurance and banking sector.

For years, both George and I hounded Peter, asking him what job Amanda did in the insurance and banking sector. But the only answer we would ever get was, 'She's in charge of schools and she's also in law court a lot.' Over time we wondered if maybe Amanda was in some kind of clandestine position within the insurance industry, and he couldn't talk about it. Eventually, we stopped asking Peter.

During our lunch meetings, the boys would both go into detail about their treatment. However, neither of them ever showed any animosity towards me for not getting liver cancer or for the fact that the hepatitis C treatment had worked for me. On the contrary, they were genuinely elated for me, which made me feel even worse.

Peter did his best to keep up with the jokes and the silly stories at every meeting, but our smiles and laughter diminished a little more on every occasion, especially after one of their regular horrible treatments. When our discussions turned to deciding whether to take up the transplant option or just give up, Peter and I were stunned to hear George say he didn't want to ever have a liver transplant. George refused to even consider the option, his decision was final, no liver transplant.

'Look, I hate pain. I'm not going to have a liver transplant. You do what you want Peter!' George explained before finishing his coffee.

'I'm just saying I'm going to give it a go, fuck it!' Peter screamed back out loud.

'That's good for you, no me. No way!' George replied not concerned at all.

From when I'd first met George, I had always been awed by his intelligence. I couldn't ever fault him with his decision-making skills and had always looked up to him for wisdom and guidance, even though I towered over him physically.

For years, I had been waiting for him to make a bad decision, but it had never happened. He had been so steadfast and consistent that I finally gave up waiting for him to be wrong about something. With his binary view of the world, if you were waiting for him to make a mistake, you would be waiting a long time.

George had always counselled me over the years without being condescending or showing any animosity when I continuously pushed myself so hard with everything. He had been genuinely concerned for me but was continually perplexed as to why I sweat the small stuff. I had my family, my work and my education and didn't need to do three jobs at the same time, especially with my condition, or take on the additional responsibility of attending disability committees, even though I was trying to better the lives of people with a disability. He genuinely worried about me, but he didn't make a fuss about it. He had always been Mr Cool-Calm-and-Collected. Nothing bothered him much and if something did, you wouldn't know unless he told you.

When he told us he wasn't going to have a liver transplant, I believed he had finally made a bad decision. I couldn't believe my ears, but instead of bouncing on it as I had envisaged doing for so long, I sat there with Peter and just listened.

I didn't need to tell him that everyone's going to be dead for such a long time so why not do whatever he could to avoid dying sooner rather than later, because he knew that. I couldn't compete with him; I wasn't on the same level.

When I asked him what I could do for him, he just smiled and asked, 'Can we continue our lunches as long as possible?'

Then when Peter replied verbatim, my heart broke. As I tried to rein in my emotions, I let Peter take over the conversation.

'Well, I've registered on the waiting list for a liver transplant!' Peter yelled without warning.

'I'm happy for you.' George sincerely answered back.

Then Peter started on George and didn't hold back but the more Peter pushed, the more George stood firm in his decision and Peter stopped just before it got ugly.

Aside from our lunches between our blood days, I had hardly had a chance to see George since our single days, but Peter had been visiting George practically every second day since taking a settlement package from his job at Telstra. Making use of his abundance of transferable skills, Peter had picked up some jobs as a coffee machine technician before stopping work for his treatment. With all this free time, the two of them had been in each other's pockets.

I think George liked the lunches so much because he had a chance to spend time with us together over lunch and no matter where we went, either Peter or I would spend time with him, and mostly at George's place, for a coffee afterwards.

George enjoyed spending time with both of us but also appreciated some time with each of us separately. It wasn't because we talked about each other behind each other's back—we did that in front of each other—it was the fact that George was the most relaxed and able to enjoy having company in his home.

I don't remember ever having anything but a wonderful time with the boys together, but I also loved those times drinking coffee with George and reminiscing with him at his place after dropping him off after lunch with Peter.

When George phoned me in late August 2015 to remind me he was well enough for me to pick him up to go to lunch with Peter at Doyle's Seafood at Watson's Bay, I was thrilled to bits and confirmed an eleven am pickup for the next day. On the following day, I wasn't too concerned with his appearance and when we got back to his place afterwards, everything was normal. I spent the entire afternoon with him as I had a day off from work and didn't need to go home in a hurry.

I left thinking George's complexion was maybe a little off but assumed it was due to him being in pain from his recent laser therapy. Both Peter and

George had agreed to use laser therapy to shrink or destroy the tumours in their livers. But by then, Peter had stopped the laser therapy and recently had a huge part of his liver removed that contained most of the tumours. This was a horrific operation and my mind had been mostly occupied with Peter.

By this stage, Peter was out of hospital and creating havoc again but was still very sore. None of us, especially Amanda, Ally and Michael, could believe that he was yelling and demanding to go home only hours after his surgery. He was so inspiring that I was secretly hoping this might have changed George's mind, but he stood his ground and refused to consider a liver transplant.

On our following blood day, Peter tried to prepare me, but I failed to conceive how far George had deteriorated in such a short period. I was devastated to hear that in one of his laser treatments, the cancer had somehow entered his bloodstream and ended up lodging on the front and back of the top of his head.

'Listen, when he turns up, try not to talk to him okay.' Peter whispered across the room.

'What the fuck are you talking about, fuck off!' I screamed back out loud.

'Okay, don't say I didn't warn you!' Peter yelled back even louder.

When he finally turned up, taking one step at a

time as if he were sleepwalking, his sister Jane always at his side helping him, I didn't recognise him. There was no way this could be my friend George!

Only days later, he was hospitalised and all we could do was to be by his side. Peter and I wasted no time in arranging for Sister Sue Shaw, our dearest nurse from the children's hospital who had faithfully looked after us as if we were her children, to visit him one last time. We did our best to warn her, so she didn't get too much of a shock, but she was magnificent.

We were all exhilarated to see George revive from his comatose-like condition when Sue stroked his face with both her hands. It was like magic. Then we were gobsmacked to see the two of them in conversation as if they were at a coffee shop. George's family were all in tears watching him with Sue, and Peter and I were relieved to see George respond to her.

'Sue, you made it.' George whispered as he stared into her eyes.

'Hello George, just like the old days, I'm here to make you better.' Sue replied with a familiar smile.

'Thank you.' He barely finished saying before his eyes closing from exhaustion.

It did not last long, but we took what we could get. Sue was appreciative of our efforts, and we took her home, talking about George's surprise awakening all the way. It had only been a few years earlier that we had tracked Sue down in Maroubra and had invited her to one of our lunches at the Eastern Suburbs Club. By the end of that day, we were all spent after thoroughly enjoying catching up with one another and totally depleted from laughing so much. There was not a dry eye among us on that occasion, so we were hopeful that Sue's visit would jolt George back to life and help him fight this nightmare.

George Lampitsi was fifty-seven when he passed away on Saturday morning on the 26th of September and he was buried on Friday, 2nd October 2015 at Rookwood Cemetery after a beautiful and emotional church service. Jane and her beautiful family as well as his family, in the midst of their tremendous loss, gave George a beautiful and most fitting send-off. Unfortunately, Sue was too frail to attend. Also, the day after George's passing, Peter received the call from the hospital informing him they had a match for his liver so he couldn't attend George's funeral as he had been admitted to hospital in preparation for his liver transplant. This robbed us both of the comfort of being there together to send off our mate. With Helen and my family at my side, I did my best for both of us.

38

MEDAL OF THE ORDER OF AUSTRALIA (OAM), 2016

While being prepped for his liver transplant surgery the day after George's passing, Peter commented that it was George who had found him a liver after passing. He said it with such conviction that both his wife Amanda and I believed it. I guess we liked to think that is how it happened. Amanda, Ally and Michael were courageous at the time, and they never left his side.

When I arrived at the hospital the day after George's funeral, I was confronted with the news from Peter's family that the transplant had been unsuccessful. Panic started to set in and just after we had lost George, things were looking dim for Peter as well.

It took me a second or two before I pushed that

thought out of my head and ushered his family to the hospital coffee shop. There, I promptly reminded them that this was Peter we were talking about.

'Hey guys, over the years, he had always managed to get out of incredibly bad health situations. I don't need to tell the three of you, do I?' I explained on death ears.

'This time is different Arthur!' Amanda replied with a quiver in her voice as Ally and Michael sat and remained quietly.

'No, I don't think it's different. Peter is Peter, it's just a different situation.

'But the first liver failed, how long can he wait until a next one is available?' Amanda was now almost paralysed in fear.

'I hear you but again it's Peter we are talking about. You watch, things will turn out great like they always do!' I said convincingly believing my every word.

Although this situation was more serious than ever, they finally agreed that Peter had made some remarkable recoveries. He had such a reputation that seasoned nursing staff would comment, 'Not even the devil wanted him.'

A week later, miraculously, the hospital notified Amanda they had a second liver for Peter. With little time to spare, they took him out of intensive care to perform a second transplant. Afterwards, we were dumbstruck to hear they had him back in intensive care because his new liver was too large, and they were unable to close him up. The bastard only weighed around fifty kilos and had he no bloody stomach. For years, at every opportunity, he'd loved to hurl fat comments at me, with his favourite being that I reminded him of Homer Simpson. I replied, 'You're so skinny because you have bloody worms, mate!'

We'd just lost George and I was in no mood to contemplate the worst-case scenario for Peter as he was bombarded day after day with complication after complication. Amanda, Ally, and Michael raised an eyebrow or two on many occasions in response to my optimism, but it didn't stop me, even though it didn't make sense with his condition deteriorating daily. With real fear in our eyes, we anxiously and painfully waited, praying for another miracle.

A week later, we got it. Once the post-operative swelling had reduced enough, the doctors were finally able to stitch him up, which kick-started his healing process. In no time, his mouth was running insults and demands again which settled everyone back down, but no one more so than Amanda and the kids.

'You should see him, he's back to being himself

thank god. This liver is working well!' Amanda said with much relief, walking out of his ward with both Ally and Michael at her side.

When the word got around that Peter was going home, people were truly shocked and couldn't believe he had done it again. We all wished George was still around to see it. Maybe it would have changed his mind but perhaps not. Either way, it would have been nice to share this with him.

A few months later, Peter was again able to attend our monthly blood days and by the following year, it was hard to tell he had even had a transplant. Eventually, he lost his yellow glow which was caused by the bilirubin formed from the breakdown of red cells in his body. The liver helps to excrete it from the body, and it was the high levels of bilirubin that had caused his jaundice, the yellowing of his skin and eyes that had been impossible to ignore. Peter laughed with a half-smirk when, at every opportunity, I said, 'Who is Homer Simpson now?'

———

My work was going through a board of management restructure at the time, and they were contemplating something unheard of—allowing people without an intellectual disability onto the SAS board of management. Many of the current board members had spent a lifetime on other management boards and were fed up with only ever being a token representative.

Although they had noticed a wonderful change within the sector of a small number of organisations had always been doing a tremendous job of having people with a disability on their board. These extraordinary people were still sceptical about having people without an intellectual disability on their board.

They talked about a cap to limit membership of the board to only two people without an intellectual disability, but it still was astonishing to me to hear them even consider the idea.

I knew then that these wonderful individuals had turned a corner. They had grown so much as a board that they felt empowered to allow others without an intellectual disability to have equal voting rights without the fear of them taking control.

This had never been a talking point at SAS, and it was something I had avoided at all costs throughout my near two decades of working in this magnificent organisation. When I heard they were seriously talking about it, I couldn't believe my ears.

As the CEO, I made it my duty to keep my opinions to myself and leave the support people, mainly people without disabilities, to provide the necessary assistance with everything to do with their decision-making process. My job was clear—to execute the decisions made by the board and report back to them the following board meeting.

I was so proud that even if it didn't eventually happen, I knew things were changing for the better and I was lucky to be a part of that process. The support people were doing a terrific job and the board members were working more cohesively than ever before.

For years, people with an intellectual disability had been sick of not being listened to and now, instead of looking back to the past, they were looking forward with incredible wisdom and knowledge. They had risen to such heights that I thought I was in the wrong board of management meeting because I couldn't recognise them all with their revolutionary thinking.

'I propose to reduce the board from a total of ten people with an intellectual disability to a maximum of six board members, only two of them without a disability.' Robert announced smiling.

'Yes, I second that.' A cascade of voices all followed one another around the entire board table.

This blew my mind because I knew their deepest and darkest fears, but there they were freely talking about it as if they have been talking about it for years.

I was left with my mouth open until someone noticed and they all started laughing. Then, without missing a beat, they all went back to talking about the restructure again. The meeting soon ended with them instructing me to arrange specific ethical and governance assistance for the board before they made their final decision.

I was overwhelmed and impressed with their decision, but what I received in the mail around the same time took me to another level of surprise. It was a letter from the Governor-General of Australia of behalf of the queen informed me I was being considered for the award of Order of Australia and would I accept it? This almost knocked me over.

I couldn't understand what was happening! I wasn't anyone of significance, so why would they give me an award like this? I reread the letter again and again, but the words didn't change, and I was bloody confused.

What made things even worse was the letter clearly stated I was not to discuss it or tell anyone about this notice because it was strictly confidential.

As soon as Helen got home, I showed her to help me make some sense of it. Although excited and thrilled for me and she commented that I was worthy of this wonderful bombshell, that didn't stop her from tearing me to bits about keeping this totally confidential until I was notified otherwise.

'I think you are missing the point Helen!' I yelled back still in shock.

'It's you that has missed the point. I told you they got the right person. You make sure you don't tell anyone about it, okay!' Helen said not daunted at all before going back to what she was doing.

I couldn't tell a soul and for weeks, I walked around like a zombie thinking that maybe they had made a mistake. Then on the 20th of May 2016, another letter came from the governor-general informing me that I had been approved for the award of the Medal of the Order of Australia (OAM), in effect from 13th June 2016. It was no mistake, because I was instructed to attend New South Wales Government House on Friday, 9th September 2016, by 2.30 p.m. for the ceremony where His Excellency General The Honourable David Hurley AC DSC (Ret'd) Governor of New South Wales would be presenting me the award, along with many others receiving awards.

The letter also stated that my award was in recognition of outstanding achievement and service to Australia and humanity. Specifically, it was to recognise my service over many years to people with a disability through a range of organisations. The governor-general, on behalf of the people of

New South Wales, thanked me for my service to the state and nation. I was astonished and humbled. In fact, I was speechless. I wondered about the upcoming ceremony and had no idea what was in store for me.

The day finally arrived and, after having our identification checked and being cleared to enter the grounds of New South Wales Government House, Helen, Jimmy, Pamela, and I took in the majestic view as we walked slowly from the car park. Once inside the building, following a second identification check, I was separated from my family and escorted to an adjacent room that was just as spectacular. I was taken to my seat, one of many regal chairs, to find waiting for me a package with my credentials, stamped with my name in gold lettering.

The entire ceremony was amazing, with all its pomp and circumstance, and having my family there to see it was magnificent. The Governor of New South Wales, who happens to now be the Governor-General of Australia, was wonderful and made each of us feel special no matter how uncoordinated or tongue-tied we were.

'Did you see that, what about that!' I asked still on a high from just receiving my medal.

'It was very impressive; you deserve it love!' Helen replied proudly.

'Congratulations, dad!' Jimmy and Pamela said

while we all made our move to the back of the grounds together.

As I walked out with my family pompously now wearing my medal, I felt a new sense of achievement the likes I never felt before and I was finally at peace with myself with owning it all now.

Afterwards, we and our families were invited for drinks and hors d'oeuvres on the grounds with magnificent Sydney Harbour views. Reality hit home for me when my family pointed out my medal, displayed for all to see, pinned on my suit jacket.

'Wow dad, you really have done well for yourself. Well done!' Jimmy burst out with much sincereness.

'Yes dad, congratulations. That is quite an achievement. We are all proud of you!' Pamela also said with a beam.

As we took in the splendid harbour and garden views, we were offered a selection of alcoholic beverages and an assortment of hors d'oeuvres. The drink calmed me down, but not for long as I bumped into the then-New South Wales Fire Commissioner, Shane Fitzsimmons. His was a familiar face as, every summer for over ten years, we had found ourselves glued to the TV for regular bushfire updates.

As boating enthusiasts and while taking occasional family holidays, the commissioner had been a vital source of information regarding bushfires, which he delivered in an expert and professional manner. I don't know how he consistently kept on doing his job without missing a beat year after year, especially with horrific summer fires that were devastating for people around the state and also for his fellow firefighters, both volunteer and regular firefighters.

'Hey dad, isn't that the fire commissioner from TV?' Jimmy yelled out as his pointed with his glass half filled with Champagne.

'Hey-yer, it is Shane!' Pamela replied now also pointing with her glass.

Shane was there as one of the VIP guests on a platform positioned right behind the New South Wales governor throughout the presentation. Meeting him in the gardens was a highlight for all of us and he was everything he appeared to be on TV and more—genuine and full of sincerity.

My family and I cherished this day and spending time among the best of the best—people who had achieved incredible things or risen to great heights within their field of expertise. I found myself rubbing shoulders with these elite individuals and the thought that we had something in common, a love and passion for helping others was very humbling.

39

TOM STILES, 2017

Things were getting unusually busy at work as years of continuous funding applications for additional programs had finally paid off and secured us some government funding to run a few short-term local disability programs. After running the programs, we went even harder the following year to secure an extension of funding for them as well as applying for funding for other programs.

Coincidentally, we happened to have Ross Lewis, the former CEO of Breakthru, a disability employment service, assisting me with a few small projects after recently retiring. Ross had also been employed for six months at SAS before I'd first started working back in 1996, and some of the original board members still knew him very well. With

Ross's experience, especially within the disability sector, he was a shoo-in, and the board took to him like a bee to pollen.

Having also secured a wonderful contractor and an expert consultant within the disability sector called Maria Katrivesis, by the time Ross came on board, our new board was ready for the extraordinary heights they all would be propelled into the following year.

The board restructure was approved and, after the training, policy, rules changes and all the interviews were over, we had two people without a disability out of a total of seven board members. The new board members were Professor Iva Strnadová from the Special Education and Disability Studies Department of the University of New South Wales and John Beard, a retired Telstra specialist, who had taken over supporting the board for his wife, Lynette Beard, after her unexpected passing a few years earlier. We were lucky to have these two extraordinary people on our board of management. In addition, the distinguished Trevor Parmenter AM, Professor Emeritus from the Sydney University, became our patron.

The remaining board members at the time were led by Robert Strike, showed they had reached that level of self-advocacy and were comfortable and secure in themselves and not afraid of anyone taking away their authority.

It was a brave new world for this extraordinary group of people with intellectual disability, and I was proud and excited for them because I know how hard it was for them to get to this level. The flow-on effect was huge for them individually, with each of them starting to achieve their personal goals, one at a time. Their confidence was showing, and the world was opening up for them as they led the way for others to follow in their footsteps.

———

It was a fifteen-year journey that led me to complete my debut novel. Writing my novel took me away from the rollercoaster ride of sensational highs and terrible lows at my work and dealing with my mortality. Dealing with people's lives as the bottom line for my organisation gave me an incredible opportunity to share the ups and downs of life with people with an intellectual disability. I experienced joy along with them as I saw their quality of life improving, but I was also there to comfort them all, including my wonderful team, when things didn't work out for the better. But the rollercoaster of work didn't compare to the continuing dilemma of dealing with my mortality.

After deciding to stop carrying the heavy load and come to terms with my mortality, I had symbolically let go of my heavy baggage when I mustered the courage to go to university at the age of forty-two. However, that was not the end of my suffering nor of my doubts and fears regarding my mortality. The doubt and fear really never left, and they manifested from time to time. Then, when the load got too heavy again, I would once more decide to let go of the burden of dying from iron overload. As the years passed, it was only too easy to convince myself that the beginning of the end was just around the corner. This was relentless and draining.

On 12 September 2017, I finally completed and self-published my debut novel. I believe that Australia is in great need of an international man of mystery who could be recognised as our very own spy agent. The first of a three-book series dream of mine about an Aussie family man with an extraordinary set of skills who becomes an international spy. Since watching Ian Fleming's James Bond movies as a young boy and after reading *Casino Royale*, I have always been intrigued as to what would take for an ordinary family man with special skills to take up this kind of work.

It wasn't until I took some creative writing workshops while studying for my master's degree in management at university that I came up with the name of my fictional hero—Tom Stiles. At the beginning of the workshops, all the participants were tasked with writing a few pages to get peer-reviewed so the facilitator of the workshops could get a better understanding of the level of our writing talent. Or in my case, the lack of it, so I thought.

In no time, I progressed from the workshops to university creative writing courses and found myself fitting in after years of feeling I was not good enough as a writer. But I guess all writers feel that way but believe me, it was true especially for me. However, I needed to work and really work hard at it, before I had something I believed in.

In fact, the professor told me to pursue my writing because he thought it would make a good book. That professor was a casual lecturer, and I didn't see him again, so I didn't pursue it at the time. But many months later, after coming across some of my writing, I started building on it to take my mind off my everyday worries.

Something incredible was happening, Tom Stiles started to evolve out of the extraordinary heights of my personal achievements that was happing in my life. The more successful I got, the more astonishing the book grew. I couldn't believe that the book became more fascinating each time I would sit down to write the next chapter.

It was so therapeutic. Writing a novel was

something I could not have ever imagined myself doing and when I realised what I was doing, I stopped immediately. Who was I kidding? I wasn't a writer. Feeling like a fraud, I packed it all away.

However, as soon as the stress of work started building up or thoughts about my mortality returned to circle around and around in my mind—they never actually ever went away—I found myself drawn back to Tom Stiles. But by the time I had finished writing the story, it was a mess and I hated it. Again, I abandoned it.

Months later, when I revisited my manuscript, I found I had developed a love/hate relationship with it. It lured me back like a siren song, but I didn't know how to fix it. It was frustrating because I needed help but the more help I got, the more unsatisfied I was with it because there was a lot I didn't understand about writing. Various people helped me, but I couldn't move forward because I lacked the fundamentals, so I decided to attend a few more writing courses at the University of Sydney.

Upon meeting part-time lecturer Christopher Cyrill for the second time, I summoned the courage to ask him if he would consider being my mentor. Christopher was at the tail end of completing his PhD, and his time was limited during this period. Also I had no idea that nearly every student had asked him that since he'd commenced facilitating these high octane and dynamic creative writing classes at the university. He politely declined, but out of frustration, I handed him the entire seventy-thousand-word manuscript of my draft manuscript.

Although, over the years, students had constantly hounded him to be their mentor, they also wanted him to do all the work for them and often would not produce a single word before approaching him. For those who did write something down, a chapter or two with a synopsis was far too short for him to take them seriously. So when I placed the entire manuscript in his hands, he looked up, surprised, and said, 'What's this?'

'It's my manuscript, is seventy thousand words enough?'

'You're kidding.' He flicked through the pages.

'No, why? Is that enough words?'

'That's fine, that's fine. Yes, that's fine, Arthur,' he repeated over and over again with a smile, still flicking through the pages.

'Look, if you become my mentor, I will work so hard for you. I will not let you down. How about it, can you give me a go?'

'Yes, yes. Can you email this to me?'

'Does this mean you'll be my mentor?' I asked taking a step back feeling totally shocked of his response starting to feel my hands sweat profusely.

'I've never had anyone put this amount of words down to show me. This is impressive, Arthur. Yes, I would love to be your mentor. Look, give me a few days and then I want you to email this to me, okay?' Chris said with a nice and pleasant change in his voice as he was still flicking through the pages.

I had no idea how impressed he was that I had a completed manuscript. He was captivated because, in all his years of teaching, no one had ever approached him with anything close to a finished manuscript. He immediately agreed to be my mentor.

I wasn't expecting this, and it took me a few days to take it all in.

After emailing my entire manuscript, the first thing he got me to do was to read Cormac McCarthy's *No Country for Old Men*. The book was incredible, but for someone who was just learning the craft, at my old age, it was also daunting.

Then, to make things worse, Chris blew my mind by condensing the first five pages of my manuscript into less than one page. A new world of writing opened up and the art of not boring the shit out of the reader was kindly and skilfully introduced to me.

It was a baptism of fire, learning about various writing techniques like reflecting and foretelling and being introduced to the phrase 'kill your darlings'. I was trying hard to understand it all, hanging on to every word and bit of advice that Chris gave me.

Ten years had passed from when I first started writing the Tom Stiles thriller until I finally met Chris. From the beginning, various people had advised me to get a good mentor, but the thought of it would leave me feeling inadequate. It took me ten years to finish the manuscript and at the end, I felt more inadequate than ever. But Chris was a gift from God and his persistence paid off for me. By the time the rewrite of my manuscript was finished and to my horror, it ended up a mere fifty thousand words. Finally, I understood what he meant by with the term of 'killing your darlings!'

Even more astonishing was Chris's suggestion to take the now complete manuscript to an editor before even considering showing it to a literary agent or publisher. No matter how much I argued that his suggestion was ridiculous because I thought the manuscript was perfect, he assured me that even professional authors get their work edited. But the secret was to get the right editor for your genre.

He was right again and after finding an exceptionally talented editor, I was horrified yet again to discover that after a copy edit, my manuscript first titled *Pivotal Velocity* was now down to forty-eight thousand words. It was 2015, and after a few days of settling back down from the shock shrinking of my work, I remembered that Ian Fleming's debut novel *Casino Royale* was only forty-two thousand words long. I was back in the game!

Buoyed by a new sense of confidence, or perhaps delusion, I went on my merry way trying to find a publisher for my manuscript, but it wasn't long before I realised I was on the road to nowhere. Two years later in 2017, and a title change from *Pivotal Velocity* to *Bang and Burn*, I self-published my Aussie James Bond style spy thriller myself and ploughed ahead, submitting it to publishers. While still waiting for a publishing contract years later, I commenced my second book titled *The Book Glasses*, the first of a two-book series.

Finishing this book in only six months and in the year of the covid-19 pandemic, didn't stop me submitting it to publishers. I didn't care, my anguish of not getting published after so many rejections over the years, my frustrations blinded me from the reality that this was the worst possible year to write a new book and having anyone review it for publication. But to my astonishment, *The Book Glasses* was successful almost immediately and it was published actually on the 9th of February 2021.

Then, to my amazement, after years of rejections *Band And Burn* was also published retitled for the final time as *Black Ops: Zulu*. *Tom Stiles*

Thrillers Book 1, on the 22nd February 2021. It was official, I'm a traditional published author now with two books published. I couldn't believe it that this happen in the same month I turned sixty years old. I'm thrilled to announced that I'm currently completing *The Book Glasses* sequel by early 2022. Soon after that, I will be starting on *Black Ops: Zulu. Tom Stiles Thrillers book 2 and 3.*

40

A FAMILY'S LOVE, 2018

One month short of fourteen years since the passing of Helen's dad, ninety-three-year-old Sotiria (Sofia) Didaskalou finally succumbed to a long battle with cancer and other age-related illnesses on 6th April 2018.

Helen's mum was an incredibly tough individual who put up with so much pain for so long, but those last years were the most difficult. Even with all the suffering and misery, when death happens no one is ready for it. Our only solace was that she was no longer in any pain. But when your mum passes, it's only a family's love that can get you through the ordeal.

Tragedy struck us again in January 2019, with the passing of Helen's beautiful sister, Voula Hatzimanolis, from cancer that originated in her kidneys. This was more difficult to deal with because she was only fifty-eight and full of life.

Voula was one of those rare souls who, if you bumped into her on a street full of people, would draw you away from the crowd to talk to you. She was a unique individual who would give you so much in return if you chose to stop and talk to her. For Voula to be taken so quickly from us without giving us any time to take a breath after the passing of her mum, it was a devastating and cruel thing for Helen to go through.

Voula's husband, Steven Hatzimanolis and their son, Angelo, were left traumatised by her rapid decline, but her death was paralysing for us all. Her death also left their daughter Sofia, and her husband Joel, trying to explain to their pre-schooler children Monica and Patrick that their grandmother will not be visiting them anymore.

Six months earlier, nineteen-year-old Angelo had commenced working part-time with me at Self Advocacy Sydney soon after completing his three months' work experience. Just when both Voula and Steve started to get some joy watching their son flourish with his first real job after school, their joy and happiness were snatched away from them unexpectedly with her passing.

No one would have been surprised, after his

mum's death, if Angelo had spiralled out of control or even lost his job during the ordeal. In fact, the opposite happened—he threw himself deep into his work, trying hard each day to make a success of himself. It stunned everyone.

Achieving full-time employment soon after wasn't a coincidence for Angelo, although he loves the work and the people he works with and for, it was because he wanted to make his mother proud of him. He knew she was watching over him.

At this especially difficult time, our souls were depleted of joy and life. We were all on autopilot, going about our daily routines. Helen was working fulltime as a fraud manager, now at another bank. Even having both Jimmy and Pamela comforting her which helped a lot with her grieving process, I was still worried for her because I knew her so well. The grief of losing both of her parents and then her much loved older sister in such a short period was a lot to take on board and I could see her pain increasing.

Worried that her pain would take her to a place where the kids and I couldn't help her anymore, I desperately wondered how I could better support her.

Then it came to me. Back in 2017, Pamela, Helen and I had taken a magnificent trip to Athens, and from there a Princess cruise to about dozen

Greek Islands. Helen had loved the whole trip. Soon afterwards, Jimmy took a work trip to Europe and stayed on for a short holiday before fitting in a quick coast-to-coast American tour on the way home. Our one-month holiday didn't leave us with much time to fit in Europe, so we pared back our trip to Greece. But perhaps a change of scenery would do Helen the world of good.

With no time to lose and with Helen screaming at me in the background, I arranged another a trip to Athens, but this time flying first to London, of course. After stopovers in London and Paris, we planned to head to Italy and embark on another Princess cruise down to Athens, stopping at as many Italian and Greek Islands as possible. Our trip was going to conclude with a ten-day stay in Athens before flying back to Sydney.

'What about your blood? See, we can't go. Now is not a good time!' Helen said taking a step back away from me.

'Hey, you know very well the trip will be only for four weeks. No different than our last trip, we talked about this. Nice try!' I replied matching her every step.

I was on a mission and finally managed to get Helen on board and agree to the trip, no matter how ridiculous and expensive she considered it. Furthermore, I had my sights set on travelling in August

2019 and made sure to lock it in before she changed her mind and insisted we wait until 2020. Lucky for us, we did, because we all know how 2020 turned out.

———

There isn't a day that I can't believe I'm still here walking around and seeing my mum, now eighty-three, and my dad, eighty-five, full of life. They still scream at each other, but other than screaming even louder and more profusely than ever, nothing has changed since I was a child. Although healthy for their age group, they are both hard of hearing, almost deaf, but refuse to believe they have a hearing problem. It is a constant battle trying to convince them to wear the hearing aids they never use.

But I am in awe of them both regarding how they go about their day-to-day life with all of their age-related difficulties, yet still full of hope and praise for all of us. Their resilience is wonderful to watch. Those long talks we once shared, often with my parents sitting my hospital bedside, when we tried to ignore the despair and hopelessness of my impending mortality still occur, but now our talks are filled with achievement and fulfilment. Who would have thought? Although it was impossible to predict that we would be still having deep conversations all these years later, thankfully the tone has improved, as I have just turned sixty years old, and my parents are both in their mid-eighties.

Most importantly, the three of us (and my brothers), are blessed that we are still here, despite thinking we were subject to a family curse. Over the years, whenever I interpreted our so-called family curse as being in any way a positive thing for me by saying that it was a blessing or I was lucky to have thal because I have met so many beautiful thals, my parents lost their minds and walked away rapidly, making the sign of the cross nonstop, hoping that God didn't hear me. Faith was their saviour and, just as they had pledged all those years ago, they have maintained an annual pilgrimage to the Greek church to give thanks that my life was spared.

Throughout our lives, no matter what they were doing or where they were, if any of their kids needed them, they were there in a flash, offering anything and everything to help us with whatever we needed or asked for. Who can ask for more than that? What made it even more special was they didn't have much to give.

My youngest brother, Angelo, after splitting up with his long-term girlfriend, moved back home some years ago and has been looking after Mum and Dad wonderfully ever since. His painting busi-ness is flourishing once again after a business partnership split put a stop to it not long after he and his girlfriend went their separate ways. After working for others for many years it's great to see him running his own painting business and controlling things again. My parents' continued good health hasn't come about by chance and the family owes Angelo a great deal for the support and love he has selflessly dedicated to Mum and Dad over the years.

My brother, Con, also had a relationship split some years ago but was fortunate to have two beautiful girls, Sofia and Angela, who we all love deeply. Having shared custody with their mother since their split, Con, and all of us, have been fortunate to watch them grow into two gorgeous teenagers. Considering that Con is a livewire with a driven personality who calls a spade a spade, his daughters have turned out marvellously. They are beautiful, courteous, smart, and full of joy and happiness. I can see in his eyes and by the way he talks about them how proud he is of them, even though he considers himself to be an absent dad because he only has them half of the time. Somehow that hasn't bothered the girls, who still adore him.

For all his daughters' lives, Con has worked as a sales manager for a prestige car brand, although he recently moved to another brand, selling both new and used cars. His girls are proud to know their dad is highly regarded and in great demand within the industry. Con's girls have given so much happiness to everyone in our family, especially their grandparents. It has been wonderful to watch how much joy they have brought my parents over the years, with sleepovers and surprise birthday parties filled with as many balloons and streamers as any kid could dare to imagine, evoking fond memories of Jimmy and Pamela's special occasions when they were young.

Even without an occasion to celebrate, when the girls are at their grandparents' house, they will often cook up a storm and have so much food left over. It's incredible to see my mother still doing those sorts of things at her age.

My oldest brother, Nick, has been living interstate for over ten years with his girlfriend, Katrina, who is an accountant. He too has a painting business that's going extremely well and even with his bad knee and hip from a soccer injury in his youth, he manages to carry on with work as if nothing was wrong.

Not surprisingly, he hasn't changed much since he was a kid. Back then, no matter whether he was at school, playing sport or even at church, he was unstoppable. Whenever he was presented with a challenge, he was hell-bent on winning no matter what! Once I saw him take out a linesman after getting a red card from a referee in a soccer match because he was sent off and couldn't finish the game. Apart from me, no one in the stadium noticed him calmly walk off the grounds, change his shirt and breezily walk over to the linesman and king-hit him. Although his efforts resulted in the longest suspension the club has ever awarded a player, it didn't bother him because all he could think about was that his chances of winning were ruined after being sent off.

No pain, distractions or any other influence can break Nick's focus if he has set his sights on achieving something. From an early age, I knew to keep out of his way. But one thing I most appreciated about my older brother was that, when we were kids, he didn't treat me any differently from my other brothers because I was weaker, even though I was older than Con and Angelo. We all experienced the wrath of Nick in equal measure, and we loved loathing him for that, even though he would most likely consider his less-than-benevolent treatment of us a compliment.

It was also good to see Helen's brothers, Nick and Con, eventually settle into their new arrangements after the passing of their mum. The passing of their mum and sister impacted them greatly and it was wonderful they chose to maintain close contact with us all and each other, especially since they were both single.

41

NOWADAYS, 2020

Since the day we were introduced to Prof. Joy Ho all that time ago, she has never dropped the ball as she's looked after us all. Just recently, she noticed a mole on my right shin and insisted I get it checked out immediately. Trying to convince her that I was satisfied it was not a problem after many recent visits to my local skin cancer sun spot specialist was futile. I hated taking time off work to visit the specialist Joy referred me to, but boy, was I glad I did. After finally convincing me to have it removed, the specialist confirmed that the mole contained a minimum of three different cancers.

Whether it's an appointment consultation or a regular blood day, Joy is always focused and directs her full attention to her patients. What has a mole got to do with thal? With Joy, you get the entire professional package and if you have something else wrong with you, she'll find it.

I could give you a long list of examples of how Joy has gone above and beyond to care for her thal patients. If any of us thals were admitted to the RPA for something other than thal, even without contacting Joy or doctors who work for her, I estimate it would take less than twenty minutes before she came and found us, even if we were still in the waiting room of RPA's emergency department.

I don't know how she does it, but Joy is everywhere at all hours of the day and night. Stories of her appearing out of nowhere when you need her the most have achieved legendary status over the years. It's quite uncanny.

For close to two decades, all the thals at the RPA have benefited from the extraordinary efforts of Professor Ho, who prefers to be called Joy.

Nowadays Joy's job title has a lot more letters after it:

Professor P. Joy Ho AM, MB.BS. (Hons), D.Phil (Oxon), FRACP, FRCPA, FFSc(RCPA) is the Senior Staff Specialist in Haematology, Director of Research and Head of the Multiple Myeloma Research Unit and Thalassemia/Haemo-globinopathy Unit at the Institute of Haematology at the Royal Prince Alfred Hospital (RPAH).

There is no doubt in our minds that Joy is heaven-sent. We have all benefited from her spectacular care and devotion. George, Peter, and I all visited each other at one time or another after being admitted into a ward at the RPA for reasons ranging from gallbladder keyhole surgery, irregular heart-related issues, liver transplants and many other problems. On every single occasion over the last twenty years, Joy has fought for us tooth and nail ensuring we had the best care and the right doctors.

When it comes to all thal-related issues, it has been comforting to have Joy at the helm of her department, leading the way with incredible professionalism and making sure things run smoothly. Her tireless efforts range from ensuring our blood expiry dates for our transfusions are consistently the best possible available, right through to regimentally checking our monthly blood cross matches to make sure all our results from iron to haemoglobin levels are all within acceptable ranges.

But, if, God forbid, Joy finds out your levels are outside the acceptable range and you were responsible for it because you hadn't been using your Desferal pump regularly or failed to comply with any of her recommendations, then you would see a different side of her. She may be shorter than all of us, but when she's unhappy with us, she's seven feet tall and as fearsome as any giant could be. It wasn't that we were scared of her, we were terrified.

'How often do you use your desferal pump Arthur?' Joy would always first ask while reading the latest notes in my files with such interest.

'Yes, I use the desferal pump.' I quickly answered back confidently then holding my breath.

'So you are using the pump?' Joy replied only after she stopped reading to focus her full attention at me.

'Yes.' I explained now holding a little less amount of air in than my last.

'Good, how many days each week?' She asked followed with a death stare and a smile.

'Four days a week.' I reluctantly let out with much relief back breathing normally.

'We agreed five but thank you for being honest, Arthur.' Joy said before going back to reading my file.

However, why we would react that way was more to do with us trying to please her rather than Joy being so scary. We older thals never received the great results the younger thals did, so we were always in catch-up mode, or that's how we felt, anyway. We never heard Joy shout or speak loudly in anger. In fact, we never noticed anything other than her beautiful and sincere heartfelt concern for all of us.

Many thals are often focused on doing what others take for granted, like having a relationship and children or seeking a future in a line of work that wouldn't suit anyone who requires regular blood transfusions. This was not so much the case for us older thals because we had an expiry date.

For the younger thals, so much of their focus is on striving to achieve personal goals, so much so that they can't hear or take in Joy's priceless information and advice. The reason why we older thals are terrified of Joy is because we know that everything she tells us to do is in our best interest. Thals have been and always will be her main priority, and nothing will distract her from focusing on a strategy that will give us the best chance of living a full and healthy life.

In our many consultations with Joy, at some time or another, all of us have not liked something she's told us to do and have pushed against it or reacted to it some way because it feels as if someone is trying to impose something on us or take away our freedom or our civil liberties. We frown and sometimes even ignore her, at our own expense.

'Hi mate, fuck I haven't seen you for years.

What are you doing in today, it's not your week!" I asked even more stunned after noticing the deep dark colour of his complexion.

'I've just got married and we are off on our honeymoon tomorrow. I'm in today to get a top up that's all.' The young thal replied with a glint in his eye.

'Congratulations, I'm so happy for you both.'

'Thanks mate.'

'Where are you and your wife heading off tomorrow?' I asked not taking my eyes off his pitch-black skin.

'Europe, my wife and I have family there. Excuse me but they are calling.' He barely finished saying before he walked out of the room.

'What was quick.' I called out just as I finally got into my chair.

'Tell me about it, apparently they want me to wait until Joy gets here before they put my IV up.' He explained more confused than me.

'Sorry, but I need to ask. Do you use your desferal pump?' I burst out unashamedly.

'Yes, why?' He told me not caring I asked.

'It doesn't look like you have, you're so dark mate. I can't even notice your olive skin at all.' I replied moving up and on the edge of my sit.

'I've been too busy to notice, really?' He answered back still not concerned.

'But aren't you a nurse?' I almost yelled trying to get his attention.

'Yes, so?' He said almost with a whisper.

'Of all people, you should know the consequences.' I barely got out before Joy burst in and whisked him out of the room.

No sooner he walked out he was back collecting his things and that's when I asked him. 'What's happening?'

'Joy wants me to be admitted into a ward today. Catch up with you next time, I've got to go.' The sullen young thal explained before quickly shooting out of the room carrying all his things.

When I found out he died the following day, it took me awhile to come to terms with it all. But what would never leave me was seeing the reaction of his gorgeous new bride, later that same day, who came in and was devastated to discover he was admitted into hospital.

I have been guilty of shooting the messenger in rebellion and ignoring Joy's advice because I was too self-centred to consider anything other than what I wanted, instead of what was in the best interest of my health. After all, without your health, you can't achieve or hang on to what you want anyway.

For the last twenty years, Joy has looked after our best interests, and words could never express my gratitude for all she has done and continues to do. Joy is truly amazing, compassionate and a brilliant professional. We all admire her both for her commitment to all the patients and families she touches every day.

Just turning sixty, I've ticked all the boxes of what I wanted to achieve and have exceeded all my expectations, and Joy has been directly responsible for allowing me to do that. Of course, she'd say it was a team effort. I'm grateful for everything Joy and her wonderful team have done for us.

———

Jimmy and Pamela are now in their early thirties, healthy and both in great, fulfilling and rewarding jobs. They have both recently had a relationship split, but you can't have everything. After graduating from school and then studying marketing, Jimmy spent the first five years of his working life in middle-management positions as an account manager in the computer electronics industry.

However, since he was a teenager, his passion has been cars, particularly BMWs, and it was inevitable he would find himself working for BMW. He started working in the used car section of the BMW dealership and all he needed to do was to sell cars to keep his dream job.

For someone who didn't know anything about selling cars and had just walked into the role of salesman, no one was more proficient or enthusiastic than Jimmy about what each model had to offer. His product knowledge was all he needed to guarantee he still had a job at the end of his probation period. After his three-month probation, he would probably have worked for free after being given a choice of any drive car to take home as part of his remuneration package.

A few years later it was no surprise to Helen and me when he took on the assistant manager role for another BMW dealer. For him, it wasn't just selling cars, because he considered BMWs to be a lifestyle, and his customers couldn't get enough of them.

Since his recent relationship split, Jimmy has been looking to buy a new apartment, townhouse or house and take advantage of the government's new home buyers' incentives. However, the coronavirus (COVID-19) pandemic hit and has temporarily halted his search. He was hoping house prices would drop by the time the pandemic was over but unfortunately, it's continuing as I finish writing my story, and the house prices have still not yet dropped.

Sadly, this current generation is dealing with a hugely inflated Sydney housing market that not even a pandemic was able to deflate. But Jimmy is undeterred—it hasn't stopped his search; it has simply slowed him down.

42

IF I COULD GO BACK IN TIME, 2020

From a young age, Pamela was a precocious child, and we knew she was going to be a lawyer. Even as a toddler she would send her mother and me to our room if we dared to challenge her wishes. Seeing her graduate with remarkable marks from university with a double degree in law and arts was the proudest moment in our lives.

Jimmy, Helen, and I were all in awe of her achievement and how she's stayed steadfastly on her path since she was a young girl, never faltering or getting distracted, especially during her six years at university. Over this period, I taught at the same university and Pam would often track me down once she heard my laughter in the corridors and we shared some priceless moments together.

'Hi honey, how did you know I was in one of the library rooms?' I said walking out with my students for our lunch break.

'I can hear you laughing on the other side of the library.' Pam screamed out before bursting into laughter herself.

'Really?' Replying trying to hide my embarrassment from my students.

'Yes, not hard spotting your laugh.' She said with her hand over her mouth.

'Do you want any money; can I give you some money?' I instinctually asked.

'You always ask me that, no!' She said with a gentle shove.

'Sorry, let's go for lunch then?'

'No, I can't because my class is starting soon. Anyway, I've had lunch. Got to go!' She announced with a giggle before heading off back to her class.

Family law was her chosen speciality but once she started working in that field at Parramatta Family Law Courts, the reality of the work was not what she expected, and an immediate career change was needed. After hearing her stories about the family courts system, neither Helen nor I argued about her change of direction.

After a few years in the legal and other various department of a prominent insurance company, Pamela got a wonderful position at one of Australia's major bank, where she found her calling in the area of risk management and she has excelled there. But Helen and I can see that her best is yet to come. A few years ago, she secured a bank loan to purchase a huge brand-new apartment in a great location and has fully furnished it. We admire the way she has managed her finances and yet is still able to live an enviable lifestyle. Of cause, she loves her BMW.

———

Amazingly, I have now got my iron overload under control with my ferritin levels hovering at an astonishing one thousand mark. In addition, I have continued to avoid having a portacath inserted into my body, to my relief. I'm still visiting my cardiologist for annual check-ups and Professor Jeremy has not yet needed to operate on my aortic aneurysm. The growth, or lack of growth, means that the aneurysm has not exceeded the maximum acceptable size which would prompt the professor to do any medical procedure. This has continued to both perplex and surprise me to this day.

Our blood days at the RPA's Thalassaemia and Haemophilia Clinic are a little less enjoyable with the absence of George. However, clinical nurse consultant Stephen Matthews who, for over fifteen years, has been the person solely responsible for putting up our monthly IVs, has been fantastic in looking after us. Steve took over from Fiona Rennison, who went to live in the UK, and his skills and professionalism have been equal to Fiona's in every way. Also, his wonderful sense of humour has made us miss her a little less and helped us cope without George. Not having to wonder if he will find a vein has been one less thing for us to worry about on our regular monthly blood days and none of us takes him for granted. Steve and his team has continued to make our regular stays much more enjoyable while never taking his eye off the ball and always being in total control and professional in every way.

Since George's passing, Peter and I have made some adjustments at our monthly blood days, but we have lost the urge to resume our regular lunches. For a while, Peter and Amanda were living the dream and taking long Greek summer holidays during our winters, staying at his sister and her beautiful family's place in sunny Sigri on the Island of Lesbos. After so many thal-related issues and two liver transplants, this was unbelievable, and I have no doubt it helped them to work through their grief over Peter's liver operations and George's passing.

With Peter's mum, who is in her late-nineties, and his younger brother Chris and his family travelling with them to Greece as well, they had a good thing going on with his clan for a while. They were able to enjoy two summers a year and completely avoid Sydney winters right up until the coronavirus pandemic messed it up for them by restricting international travel.

Helen and I finally ended up going on our European holiday in 2019, and we were glad we didn't postpone it until 2020 or it would have been cancelled due to the pandemic. It was a dream come true. We visited London and Paris and saw some of Italy, then cruised on the Emerald Princess from Rome to Athens, stopping off at as many Italian and Greek Islands along the way as possible. Furthermore, flying from Sydney to London and then from Athens to Sydney was something special considering what's happening now. Who knows when anyone will next be able to take a holiday overseas the way we did, or anywhere locally for that matter?

The coronavirus pandemic has frightened all of us, especially the people in my parents' age group. Also, people like thals, who are more susceptible than those without any underlying medical-related issues, are panicking and have additional worries to deal with. However, having made it through the AIDS virus in the eighties, with the devastation it caused the haemophilic population, we have managed to follow expert advice and stayed focused and safe, avoiding any causalities in my group so far. It's a long road, but we need to follow it until we all have the vaccine. We all know what we need to do, and we have to accept it unconditionally.

————

If I could go back in time, I would. Yes, I would go back to those Three Musketeers days and take all the heartache away from that scared little boy who had been bluntly informed he was going to die young. I would tell him to not be afraid and that all he needed to do was focus on living, one day at a time, and to slowly chip away at the most important goals he had and not discard them. In reality, that's all any of us can do when confronted with the uncertainties of life—focus on living one day at a time!

As a twelve-year-old with a comic book in hand and a wonderful imagination, Iron Man comics saved my life because I was Iron Boy, at a time I needed a hero to rescue me when the odds were stacked against me. But as an adult, I came to realise my heroes are actually blood donors. They may not wear capes or have superhuman abilities. They may not battle diabolical villains and overcome evil henchmen and make awesome sound effects like wham! Kapow! Boom! In fact, most blood donors are ordinary people who do an extraordinarily wonderful thing—they donate their blood.

Every day since I discovered who my heroes truly are, I have tried to live my life the best way I could and to fully appreciate that precious gift of blood donated freely by people I don't know. Donated blood is essential to save the lives of so many people every day—from accident victims to those having surgeries and to people like me and my thal friends, who need regular transfusions to simply survive.

To date, I'm horrified to realise I have had over 700 blood transfusions and 2,200 donated blood packs since birth just to stay alive. From this, plus the regular use of my Desferal pump, I have endured over 8,600 needle sticks, and counting.

Back in the early 1980s, George, Sev, Peter and I were all in our early twenties when we first were introduced to this incredible new medication called Desferal (Desferrioxamine). At the time, we had no idea what a miracle treatment it was... if you used it, of course!

Even though we agreed to start this new treatment, all we took in from what the doctors said was that the damage had already been done to our bodies and we could hope for the best and for a few precious more years. It's a testament to this wonderful drug that if you use it, it really does work. Patients introduced to it from an early age now live long and normal lives, which is something we never expected to do.

These days, apart from the Desferal (Desferrioxamine) pump, there are two other options available to thals and they both come in tablet form regarding chelation therapy. They are both chelating agents. Deferiprone is taken as a tablet or liquid three times a day. It's sometimes used alongside the Desferal pump to reduce the number of infusions. Deferasirox is taken once a day as a tablet that you dissolve in a drink. These options and others have been around for over a decade now and remain bittersweet for me. Bitter, because after waiting for a lifetime for them to be available, my body rejected both treatments after only two weeks of use. Yet, also sweet, knowing that many thals around the world will live long lives and never need to use the Desferal pump.

It has been a challenge, especially during the COVID-19 pandemic, to find out if there is anyone older than Peter and me with our thal in Australia. I was unable to find anyone older, but Desferrioxamine was developed in the late '60s, so all I know is that there would be very few, if any, thals older than Peter and me still alive. My thal mates and I were the first generation of thals to live beyond our early twenties, and that was only due to this drug called Desferrioxamine. If it wasn't for the generosity of the wonderful people around the world who donate blood, people like Peter and me would not be here, nor would our many younger thal friends and others like us.

I don't like to repeat myself, but I feel in this case I must. People around the world donate blood for many reasons, but whatever they are, everyone I know who receives this precious gift is grateful because it has given us a chance to live, and we have all done something with our lives. Hopefully, I have been able to and will continue to give back in a way to show people who donate blood where their precious gift ends up.

There is no way I can truly thank the hundreds and hundreds of blood donors who have kept me alive through their gift of blood. They have allowed me to live. Without them, I would not have survived my early childhood, let alone lived a long and full life. Without them, I would have never married my beautiful wife and had two beautiful children. From the bottom of my heart, I thank all blood donors in Australia and around the world for your precious gift. It means everything to me and my family and my thal friends too.

This book is the story of my life so far, a story which would have not continued past the first chapter without the people who donated the blood that I needed to survive. So this book, and my life story, is a thank you to everyone who has or does donate blood. Thank you for your precious gift.

Blood donors gave me life, but Helen encouraged me to live. I couldn't have done all of this without my beautiful wife Helen who has been my muse and my every breath. My wife has been my conscience, my sanity, and my inspiration. I love her so much. Helen is the light in my darkness. From the bottom of my heart, I would like to thank her for her love and devotion and the steadfast support she has shown me over the years.

Watching Helen be such a wonderful mother to Jimmy and Pamela, and balance motherhood and her professional work life, inspired me to overcome my despair. I am blessed to have the love of Helen, Jimmy and Pamela, and their love forever erased the question I had faced every New Year's Day since I was a child: will this year be my last? They taught me to live, to love without reservation, to plan for the future, and to hope.

Blessed to have my parents still with us, I'm forever grateful to them both for their unconditional love, support and devotion for faultlessly navigating me on this incredible path. It was a result of them both taking small steps that I was able to run and jump faster and as high as I did throughout my life...thank you so much for everything mum and dad!

———

It's inspiring to see the many young thals I've been fortunate to come across over the years, many achieving prominent positions, becoming parents, and accomplishing great things. The stories I hear of their triumphs and tribulations are encouraging. I admire them all because they don't have the 'stunned deer in the headlights' stare that I had at a time of uncertainty and trepidation.

They bring to mind a quote by Oprah Winfrey: 'The more you praise and celebrate your life, the more there is in life to celebrate.'

Dear reader,

We hope you enjoyed reading *Iron Boy*. Please take a moment to leave a review, even if it's a short one. Your opinion is important to us.

Images related to this book are available for viewing at https://arthurbozikas.com/books/iron-boy/

Discover more books by Arthur Bozikas at https://www.nextchapter.pub/authors/arthur-bozikas

Want to know when one of our books is free or discounted? Join the newsletter at http://eepurl.com/bqqB3H

Best regards,
Arthur Bozikas and the Next Chapter Team

ARTHUR BOZIKAS OAM, JP(NSW), FAIM, FGIA

Fellow of the Governance Institute of Australia.

Fellow of the Australian Institute of Management.

Bachelor of Adults Education, University of Western Sydney, 2005

Master of Management, University of Western Sydney, 2006.

Justice of the Peace appointed 30th May 2007.

Graduate Certificate of Research Studies, University of Western Sydney, 2009

Certificate IV in Assessment and Workplace Training: (TAE40110), 2011

CEO of Self Advocacy Sydney (SAS) from 1996 to 2007, after a five years' break, was back in January 2012, current. SAS has been federally funded since 1986 to educate, support, train and provide information to people with intellectual disability develop their self-advocacy skills, improving their overall quality of life.

As Chief Executive Officer of Self Advocacy Sydney, I am proud to lead a team that supports people with intellectual disabilities to speak up for themselves and achieve their self-advocacy goals and work towards SAS' mission of creating a community where all people with intellectual disability are valued. I have more than 25 years' experience in service delivery, advocacy, policy development and management. I enjoy working alongside SAS members, striving together to build a more inclusive and welcoming community.

Casual Teacher at WSI TAFE College, 2004 to 2011, teaching business and management certificates and diplomas.

Casual Academic (Staff) at University of Western Sydney, 2005 to 2011, delivering a range of business and management units to undergrads.

Blacktown City Council Disability Access Advisory sub-committee Board Member 2004 – 2007, 2012 – 2017.
    Blacktown City Council CBD sub-committee Board Member 2004 – 2007, 2017 to current.
    Disability Advocacy Network Australia (DANA) Board of Director 2014 – 2018

Received the Joan Timperley Award for Services to People with Disability awarded by Councillor Leo Kelly OAM, Mayor of Blacktown City Council, for outstanding contributions to People with Disability on the third of December 2013 at the International Day of People with Disability at Blacktown City Council.

Honoured to receive a Medal of the Order of Australia on the Queen's Honours List on the 13th of June, 2016, for service to people with a disability through a range of organisations. Awarded by His Excellency General, The Honourable David Hurley AC DSC (Retd).

CPSIA information can be obtained
at www.ICGtesting.com
Printed in the USA
LVHW030747111121
702999LV00006B/101

9 781006 422638